Mathematics in Berlin

Editors
H. G. W. Begehr
H. Koch
J. Kramer
N. Schappacher
E.-J. Thiele

on behalf of the
Berliner Mathematische Gesellschaft

Birkhäuser Verlag
Berlin · Basel · Boston

Editors:

Prof. Heinrich Begehr
Fachbereich Mathematik u. Informatik
Mathematisches Institut (WE 1)
Freie Universität Berlin
Arnimallee 3
D-14195 Berlin

Helmut Koch and Jürg Kramer
Humboldt Universität zu Berlin
Institut für Mathematik
Jägerstr. 10–11
D-10117 Berlin

Norbert Schappacher
U.F.R. de mathématique et d'informatique
Université Louis Pasteur
7, rue René Descartes
F-67084 Strasbourg Cedex

Ernst-Jochen Thiele
Technische Universität Berlin
Fachbereich Mathematik
Strasse des 17. Juni 136
D-10623 Berlin

1991 Mathematics Subject Classification 01 A73

A CIP catalogue record for this book is available from the Library of Congress, Washington D.C., USA

Deutsche Bibliothek Cataloging-in-Publication Data
Mathematics in Berlin / H. Begehr ... ed. - Basel ; Boston ; Berlin :
Birkhäuser, 1998
ISBN 3-7643-5943-9 (Basel ...)
ISBN 0-8176-5943-9 (Boston)

Printed on acid-free paper produced of chlorine-free pulp. TCF ∞
Cover design: Markus Etterich, Basel
Printed in Germany
ISBN 3-7643-5943-9
ISBN 0-8176-5943-9

9 8 7 6 5 4 3 2 1

To Kurt-R. Biermann,

the father of the historiography of Berlin mathematics.

Contents

Preface

This little book is conceived as a service to mathematicians attending the 1998 International Congress of Mathematicians in Berlin. It presents a comprehensive, condensed overview of mathematical activity in Berlin, from Leibniz almost to the present day (without, however, including biographies of living mathematicians). Since many towering figures in mathematical history worked in Berlin, most of the chapters of this book are concise biographies. These are held together by a few survey articles presenting the overall development of entire periods of scientific life at Berlin.

Overlaps between various chapters and differences in style between the chapters were inevitable, but sometimes this provided opportunities to show different aspects of a single historical event – for instance, the Kronecker–Weierstrass controversy.

The book aims at readability rather than scholarly completeness. There are no footnotes, only references to the individual bibliographies of each chapter. Still, we do hope that the texts brought together here, and written by the various authors for this volume, constitute a solid introduction to the history of Berlin mathematics.

One cannot write about this history without constantly paying tribute to the fundamental and extensive work of Kurt-R. Biermann. This work is most visible in (although not at all exhausted by) his book *Die Mathematik und ihre Dozenten an der Berliner Universität 1810–1933, Stationen auf dem Wege eines mathematischen Zentrums von Weltgeltung*, Akademie Verlag, Berlin 1988, which is the standard reference for the subject, and at least a latent reference for all chapters of this book. We dedicate our book to Kurt-R. Biermann, the father of the historiography of Berlin mathematics.

The table of contents will guide the reader in detail. Let us simply indicate here the main periods of mathematical development in Berlin:

The Academy, 1700–1810. Leibniz was the initiator and first president of the Berlin Academy. Afterwards Euler worked there for 25, and Lagrange for 21 years.

First decades of the University, 1810–appr. 1850. This period began with the opening of the University. The budding mathematical life was strongly marked by non-mathematicians such as Alexander von Humboldt and A.L. Crelle. Outstanding mathematicians at the time were Abel, Dirichlet, Eisenstein, Jacobi, and Steiner.

The Golden Age, appr. 1850–1891. The outstanding period in the history of mathematics in Berlin is embodied by the triumvirate Kummer, Weierstrass, and Kronecker. Young mathematicians came to Berlin from everywhere to study here. For the first time, the mathematical courses were structured in systematic cycles, allowing students to take all the basic courses in a few years.

Interim, 1892–1918. A period of relative decline during which Göttingen clearly took the lead. Important figures for this period were L. Fuchs, F.G. Frobenius, F. Schottky, H.A. Schwarz, and the young E. Landau as a lecturer.

A new start, 1918–1932. E. Schmidt, I. Schur, L. Bieberbach (before taking the Nazi turn) and R. von Mises created a stimulating mathematical center.

The Nazi period, 1933–1945. Berlin mathematics under Hitler was marked by a radical politization of the discipline, the so-called Deutsche *Mathematik*, a Nazi movement (and a journal) whose Berlin exponents included, in particular, L. Bieberbach, O. Teichmüller, and T. Vahlen.

Post–World War II, 1945–1953. During the first years of the Cold War, a strong center of pure mathematics briefly flourished at the university in (East) Berlin, around E. Schmidt, H.L. Schmid and H. Hasse.

Separate histories, 1948–1990. During the existence of the German Democratic Republic the development at the university and the academy in East Berlin was increasingly independent of the evolution at the *Freie Universität* and the *Technische Hochschule* in the western part of the city.

Since the reunification, 1990. Hopes for a new mathematical flowering continue despite financial difficulties.

One topic that is only very incompletely covered in this book is the history of mathematical journals that were founded in or directed from Berlin. In particular, review journals played an important role in the mathematical life of this city: Berlin is the birthplace of the first mathematical review journal, the *Jahrbuch*, and the offices of the *Zentralblatt* were transferred here in 1940 and amalgamated with those for the *Jahrbuch*. Below is the list of mathematical journals that were founded in Berlin:

Journal für die reine und angewandte Mathematik ("*Crelle*"), founded in 1826
Jahrbuch für die Fortschritte der Mathematik, 1870–1942
Sitzungsberichte der Berliner Mathematischen Gesellschaft, founded in 1901
Mathematische Zeitschrift, founded in 1918
Zeitschrift für angewandte Mathematik und Mechanik, founded in 1921
Schriften des Mathematischen Instituts und des Instituts für angewandte Mathematik der Universität Berlin, 1932–1940/41
Deutsche Mathematik, 1936–1943
Mathematische Nachrichten, founded in 1948
Zeitschrift für mathematische Logik und Grundlagen der Mathematik, founded in 1955
Biometrische Zeitschrift, founded in 1959
Annals of Global Analysis and Geometry, founded in 1983.

We gratefully acknowledge the expert technical and editorial help that was unfailingly provided by Mrs. C. Dobers, Mr. G. Hein, Mr. W. Rietdorf and Mrs. B. Wüst. Thanks are due to Mr. T. Hintermann from Birkhäuser Verlag, Basel, for the excellent cooperation in producing this book with minimal delay.

We hope that this small volume will interest readers in the mathematical past and present of the German capital, and continue to serve as a reliable introduction to further reading beyond the 1998 ICM.

Berlin, March 1998 The Editors

Mathematics at the Prussian Academy of Sciences 1700–1810

Eberhard Knobloch

Long before Berlin received its own university in 1810, there had already been first-class mathematicians in the city and eminent mathematical research had been done thanks to the Berlin Academy. The foundation deed of the Berlin-Brandenburg Society of Sciences, today's Berlin-Brandenburg Academy of Sciences was signed by the Elector Friedrich III of Brandenburg (1657–1713) on 11 July 1700, his birthday. On 12 July, one day later, the universal genius Gottfried Wilhelm Leibniz (1646–1716), born in Leipzig, was appointed President of the new Society. He was in the employment, in actual fact, of the Elector Georg Ludwig (1660–1727) of Hanover, who had reigned since 1698 and who, in 1714, was to become the King of England. However, between the years 1700 and 1711, Leibniz spent time in Berlin serving ten terms of office as President, for a total of three years. He never again returned to Berlin after 1711.

At that time Leibniz was one of the most famous scientists of Europe and, with Isaac Newton (1743–1827), the most important mathematician of his time. Even though he did not live in Berlin much of the time, he determined the intellectual life of the Society. Twelve of the 61 contributions to the first volume of the *Miscellanea Berolinensia*, the journal of the Society, were written by him. This volume appeared in 1710 and contained three mathematical treatises by Leibniz. They concerned:

1. His index notation by means of fictive numbers. This notation played a crucial role in his studies of determinant theory, which he founded.

2. The remarkable analogy between the powers of the polynomial and the differentials of a product. Joseph Louis Lagrange (1736–1813) continued to study this analogy in 1774 during his sojourn in Berlin.

3. An application of the differential calculus to a tangent problem.

Among the other mathematical articles published by Leibniz during the last sixteen years of his life, many concerned integration theory (partial fraction decomposition), divergent series, differential calculus and binary arithmetic.

Only twenty-five years after Leibniz's death, mathematics at the Berlin Academy again reached an internationally significant, high scientific niveau, which it retained for nearly half a century. This was to a great extent due to three Swiss mathematicians and one Italian mathematician who lived and worked as Academy members during a certain period of their lives in Berlin: Leonhard Euler, from 1741 until 1766; Johann III Bernoulli, from 1764 until his death in 1807; Johann Heinrich Lambert, from 1764 until his death in 1777; and Giuseppe Lodovico

Gottfried Wilhelm Leibniz

Lagrangia, better known under his French name Joseph Louis Lagrange, from 1766 until 1787.

Euler (1707–1783) was born in Basel. He was already a famous mathematician and member of the St. Petersburg Academy when in 1741 the Prussian King Friedrich II asked him to join the Berlin Academy. The King had decided to reorganize this institution, and Euler was one of those invited to help to do this. Euler accepted the offer. On 25 July 1741, he arrived in Berlin. He became the head of the Academy's observatory in the same year. In 1743 the reorganized Academy was called *Académie Royale des Sciences et Belles Lettres*. Euler participated in its first session, which took place in January 1744. In 1746 the French mathematician and physicist Pierre Louis Moreau de Maupertuis (1698–1759) was appointed President of the Academy by the King; Euler became Director of the mathematics section. After Maupertuis's death in 1759, Euler was in charge of the Academy, but without being appointed its President.

During the Berlin period Euler wrote 380 works of which about 275 were published. He regularly sent articles to the St. Petersburg Academy, mainly in pure mathematics such as number theory, and participated in the teaching of Russian scientists: S.K. Kotelnikov, S.Y. Rumovski, and M. Sofronov studied in Berlin under his supervision for several years. Euler was "the golden bridge between the two academies," as Eduard Winter put it (see [6], p. 60).

Joseph Louis Lagrange

Eight voluminous monographs were among the publications of his Berlin period, the first of which was his book on the calculus of variations (1744), entitled *Method of finding curves having a property to highest or lowest degree or solution of the isoperimetric problem taken in the broadest possible sense.* Carathéodory (1873–1950) called it one of the most beautiful works which had ever been written in mathematics (see [3], p. IX). In it Euler still uses the geometrical polygonal method, and gives no method which is independent of the geometrical solution. Such an algorithm, such a non-geometric method, was later developed by his successor Lagrange in 1755. Euler at once adopted the Lagrangian formalism, frankly appreciating the merits of its inventor. Especially interesting in Euler's book are its two appendices. The first deals with elastic curves and leads to the solution of an eigenvalue problem. The second appendix contains a mathematically correct form of the principle of least action. Nine further articles followed dealing with this famous principle.

The quarrels about this principle between Maupertuis, the King, and Euler on the one side, and Samuel Koenig (1712–1757) and Voltaire (1694–1778) on the other resulted in a catastrophe: Maupertuis left Berlin in 1753, at first temporarily, soon definitely, though he remained President of the Academy until his death in 1759.

Also in 1744, Euler published his fundamental book on the calculation of orbits, that is, his *Theory of the motion of planets and comets*. In 1745 he pub-

Leonhard Euler

lished *New principles of artillery*, and in 1748 the *Introduction into the analysis of the infinites*. The latter consists of two volumes and paves the way for the two textbooks on the differential (1755) and integral calculus (1768). At the beginning, a function is defined as an analytical expression. The first volume explains the calculation with infinite series and products and the relative representations of transcendent functions. The second volume concerns analytic and differential geometry.

In 1749, Euler's treatise on ship building and navigation appeared, in 1753 his first theory of the lunar motion. The last lengthy monograph which was published during his Berlin period was his treatise on the mechanics of solids (1765). His famous *Lessons to a princess of Germany on diverse subjects of physics and philosophy*, originated in lessons given by Euler between 1760 and 1762 to a relative of the Prussian King, but they were published only in 1768 in St. Petersburg, after he had returned to the Russian Academy. However, the principal parts of the three volumes of his work on integral calculus were finished in Berlin.

Euler was a mathematician, in the true sense of the word in those days; that is, he contributed to all branches of pure and applied mathematics, including celestial and rational mechanics, and most of all, to mathematical analysis. Some special accomplishments of the Berlin period should be mentioned. In number theory he proved by three methods Fermat's little theorem (1741, 1761, 1763) with the outcome of the theory of residues modulo n (1760). He was the first to use analytic methods in this mathematical discipline and thus discovered important results concerning the zeta-function (1749). He expressed for the first time a piecewise algebraic function by a Fourier-series and extensively dealt with divergent series, introducing a new notion of sum (1755). He elaborated a complete theory of logarithms of complex numbers (1751). He developed solution methods of differential equations. He made substantial advances in the study of curvature of surfaces (1763). He gave a new exposition of point mechanics (1765).

From 1753 to 1755 Euler elaborated in detail an analytical theory of fluid mechanics. His first lunar theory enabled Tobias Mayer (1723–1762) to compile lunar tables which considerably improved the calculation of the position of the moon and thus of the longitude of a ship.

In addition to this tremendous scientific activity, Euler successfully participated in many prize competitions of the Parisian Academy of Sciences (he received twelve awards and had numerous administrative duties). At times he was charged by the King with practical activities, as for instance, canal building or the organization of state lotteries.

In view of the increasing differences with the King, he decided to leave Berlin and to accept the invitation of the Russian Empress, Katharina II, to return to St. Petersburg. On 9 June 1766, Euler left the city of Berlin where he had spent 25 years.

Two years before his departure, a Swiss countryman entered the scene. Lambert (1728–1777) was born in Mülhausen, which belonged in those days to Switzerland, and later on mostly – as today – to France (Mulhouse). He acquired all of his scientific training and substantial scholarship by self-instruction. After several appointments in Germany and in Switzerland, he came to Berlin in 1764 and became member of the Academy one year later. As a member of its Physical Class he produced more than 150 works for publication, which covered philosophy, natural science, and mathematics. Already in 1761, he had proved the irrationality of π and e. His largest mathematical publication, the four volumes of his *Contributions to the use of mathematics and its application* appeared after his arrival in Berlin (1765–1772). They comprehend a collection of writings on various mathematical topics, among them being a *Practical geometry* and *Theory of reliability*. Therein he continued his elaboration of a general error theory. He was indeed the first founder of such a theory. In 1766, he wrote his *Theory of parallel lines*, which was published only posthumously in 1786. There he studied the famous parallel problem, using a quadrilateral with three right angles and considering the possibilities of the fourth angle being right, obtuse, and acute. The acute angle hypothesis led him to introduce the hyperbolic functions. Hence he plays an important role in the prehistory of non-Euclidean geometry. The same is true of the theory of map construction, that is, of cartography. In his *Notes and additions on designing land maps and maps of the heavens*, published in 1772, he obtained the formulas for a conformal mapping of the sphere on the plane in full generality. Two years later, the second edition of his second book appeared, namely *The free perspec-*

tive, which is a masterpiece in descriptive geometry, containing many valuable geometrical discoveries.

In 1764, when Lambert came to Berlin, another young Swiss scholar was invited by Friedrich II to reorganize the astronomical observatory at the Berlin Academy: Johann III Bernoulli (1744–1807). This member of the Bernoulli family was the most successful of the sons of Johann II Bernoulli (1710–1790). His treatises, which were quite important, concerend astronomy, not mathematics. He translated Euler's *Algebra* into French. With Karl Friedrich Hindenburg (1741–1808) he published the *Leipziger Magazin für reine und angewandte Mathematik* from 1776 until 1789 and edited Lambert's *Theory of parallel lines* in 1786. In 1791 he became Director of the Mathematical Class as successor to Jean de Castillon (1708–1791). After his death no new director was appointed.

Lambert and Johann III Bernoulli were Euler's colleagues during Euler's last two years in Berlin. When Euler left Berlin in 1766 in order to join the St. Petersburg Academy, Friedrich II succeeded in inviting Lagrange (1736–1813) to become Euler's successor of equal rank. Lagrange accepted the King's offer thanks to Jean le Rond d'Alembert's (1717–1783) advice.

These two outstanding mathematicians, Euler and Lagrange, knew each other through their correspondence which had been initiated by Lagrange in 1754. Both scientists were well acquainted with the scientific publications and interests of the other. Lagrange very often took Euler's mathematical ideas as his own basis of thinking. This applies to number theory, algebra, infinitesimal analysis, mathematical physics, celestial mechanics, etc. The styles were very different: Euler's was more detailed, but more comprehensive; Lagrange was more concise and rigorous.

On 21 August 1766, Lagrange left Turin where he had taught mechanics and integral calculus at the Artillery School. He reached Berlin on 27 October. On 6 November he was named Director of the Mathematics Section of the Berlin Academy and quickly became acquainted with Lambert and Johann III Bernoulli. Regarding celestial mechanics, Lagrange took up Euler's studies of special cases of the three-body problem, after his arrival in Berlin:

1. He dealt with the problem of two gravitational centers which act on a third body (the two papers were read in 1768).

2. He studied the case when the three bodies move on a fixed straight line (the collinear configuration, in which the ratios of the distances between the three bodies remain constant) in his prize essay on the three-body problem of 1772. For this he was awarded the prize of the French Academy, together with Euler, who had submitted his Second Lunar Theory. Lagrange considered the collinear as well as the triangular configuration. Contrary to his own assertion, he did not use a new method, but applied Euler's method to a more general case than Euler had done.

The main portion of Lagrange's arithmetical works is concentrated in the first ten years of his stay in Berlin. They include his studies of diophantine equations of first and second degree, especially the solvability of Fermat's equation

$$x^2 - ay^2 = 1.$$

In the Berlin *Mémoires* of 1770, he set forth the first demonstration that every natural integer is the sum of at most four perfect squares. Other related treatises concerned prime number theory, continued fractions, quadratic forms, and difficult problems in indeterminate analysis.

A series of papers dealing with algebraic analysis appeared during these first ten years. The culmination of this research in the theory of equations was a memoir read in 1771: "Reflection on the algebraic resolution of equations". There the concept of the permutation group appeared, although the term itself was not used. Of course, Lagrange gained his European reputation by his works in infinitesimal analysis. In Berlin, he made a systematic study of singular solutions and their connection with a general solution of ordinary differential equations. He introduced non-constant coefficients of such equations, which led to the concept of adjoint equation. The method of variation of parameters for a single ordinary differential equation was applied by Lagrange to the n-th order equation

$$Py + Qy' + Ry'' + \ldots + Vy^{(n)} = X,$$

where $X, P, Q, R, \ldots$ are functions of x.

His contribution to the calculus of probabilities, read before the Berlin Academy, was partly inspired by Pierre Simon de Laplace (1749–1827). The last famous major work he elaborated while still in Berlin was his *Treatise on analytic mechanics*. It appeared in 1788 in Paris, one year after he left Berlin. After the death of his wife in August 1783, and of Friedrich II in August 1786, the princess of Italy and the French government competed in attracting Lagrange to their courts and to Paris, respectively. Lagrange left Berlin on 18 May 1784, for Paris. The decline of Berlin mathematics was inevitable. Only after several decades, and indeed only after the foundation of the Berlin University in 1810, did it again reach the high scientific level which had been created by Euler and Lagrange during the 18th century.

References

[1] *Borgato, M.Th., Pepe, L.*, Lagrange, Appunti per una biografia scientifica, Torino, 1990.

[2] *Brather, H.-St. (ed.)*, Leibniz und seine Akademie, Ausgewählte Quellen zur Geschichte der Berliner Sozietät der Wissenschaften 1697–1716, Berlin, 1993.

[3] *Carathéodory, C.*, Introduction to: Leonhard Euler, Opera omnia, vol. I, 24, Zürich, 1952.

[4] Leonhard Euler 1707–1783, Beiträge zu Leben und Werk, Gedenkband des Kantons Basel-Stadt, Basel, 1983.

[5] *Steck, M.*, Bibliographia Lambertiana, Ein Führer durch das gedruckte und ungedruckte Schrifttum und den wissenschaftlichen Briefwechsel von Johann Heinrich Lambert 1728–1777, Neudruck mit Nachträgen und Ergänzungen, Hildesheim, 1970.

[6] *Thiele, R.*, Leonhard Euler, Leipzig, 1982.

[7] *Giacardi, L., Roero, C.S., Viola, C.*, Sui contributi di Lagrange alla teoria dei numeri, Rendiconti del Seminario Matematico Università e Politecnico Torino **53** (1995), 151–181.

Mathematics in Berlin, 1810–1933

David E. Rowe

When Berlin University opened in 1810, Prussia stood under French occupation as Napoléon's *grande armée* swept across Europe. Five years later, the Congress of Vienna sought to restabilise European relations by largely restoring the old order, thereby leaving a major question mark concerning the political constructs that were to constitute the modern nation states within the Continent's centre. During the ensuing period, culminating in the dramatic events of 1848–49, students and other intellectuals crusaded for democratic reforms while governments cracked down on political agitation, beginning with the Carlsbad decrees of 1819. Hopes ran high in liberal university circles that Prussia, the most powerful and in many respects one of the more progressive of the German states, would take the lead in uniting Germany into a constitutional monarchy along lines similar to the French and British models. But after King Frederick William IV refused to accept the crown offered him by the Frankfurt Parliament in 1849, events soon took a sharp turn toward confrontation, and liberal dreams quickly dissolved after the hopelessness of military resistance became apparent. Fifteen years later, a younger generation would begin to discount the old ideals of the 1848ers in a world shaped by the "new" ethos of Bismarck's *Realpolitik*, the politics of "blood and iron."

Back in 1810, the world of higher mathematics, too, was dominated by France. Indeed, a sampling of Europe's mathematical elite would reveal that more brilliant minds were residing within the walls of Paris than throughout the vast expanse of terrain outside them. The era of the great international academies – when Christian Huygens adorned the Paris Academy and Frederick II preferred having Lagrange rather than the *bieder* Euler in Berlin – was now a thing of the past. The modern nation state, whatever form it might take, carried with it as a matter of course the notion of national scientific communities, and within these mathematics would presumably find its niche alongside its sister disciplines, astronomy and physics. Mathematics had never played a particularly prominent role in the curriculum of the European universities, whereas the Paris Ecole Polytechnique was turning out throngs of first-rate mathematicians as a mere by-product of its curriculum for training engineers. To any informed contemporary observer the prediction that within a half century the German mathematicians would overtake their French counterparts and that the newly founded *Friedrich-Wilhelms-Universität* in Berlin would replace the Ecole Polytechnique as the foremost training ground for young mathematicians would surely have sounded ludicrous.

True, Gauss had already established himself among the luminaries of the age, but his part in elevating mathematics within the German universities had always been modest. By his own choice, Gauss stood outside the mainstream institutional developments that led to a major burst of high-level mathematical activity particularly in the northern German states. Despite enticing offers from Berlin

Carl Gustav Jacob Jacobi

and the entreaties of his friend Alexander von Humboldt, he resolved to remain in Göttingen, thereby cementing the mathematical reputation of its prestigious Hanoverian university. Gauss's scientific network, which overlapped with Humboldt's own, was largely comprised of professional astronomers. Younger mathematicians, including Abel and Jacobi, regarded him as nearly unapproachable and, with the notable exception of Gotthold Eisenstein, he rarely uttered more than the briefest of praise for their work. Those who attended Gauss's courses in Göttingen gained the pleasure of hearing him explain error theory, least squares, and related mathematical methods of fundamental importance to those working in astronomy and geodesy, but he never offered lectures on the topics that have made his name so famous today (within the vast literature on Gauss, one might begin with [9].)

Practical mathematics of a similar ilk was taught by the professors in Berlin, men with such eminently forgettable names as J.G. Tralles, L. Ideler, E.H. Dirksen, and M. Ohm. Deservedly or not, their reputations died with them as representatives of a style of mathematics whose principal *raison d'être* resided in service to the natural sciences. Thus, before the late 1820s, no one who yearned to make a name for himself in higher mathematics would have been well advised to study in Berlin (or at any other German university for that matter). If young Carl Gustav Jacob Jacobi did not already know this when he enrolled in 1821, it took him little time to discover that the Berlin mathematicians could offer him nothing to sustain his interests (on Jacobi's life and work, see [11]). With no apparent regrets, Jacobi turned to classical philology instead. Having already digested Euler's *Introductio ad analysin infinitorum* before entering the university at age sixteen, he could afford to pursue his mathematical interests through private reading, just as Gauss had done as a student in Göttingen. In this era of the "self-made" mathematician, Jacobi took the next step up his quaint and curious career ladder four years later by simply presenting his credentials to the Berlin faculty, whereupon he was promptly appointed *Privatdozent* (a position that normally carried no salary, though it did entitle the young scholar to collect fees paid directly to him by his auditors). One year later, Jacobi accepted the Prussian Ministry's offer of a post in Königsberg, from which station he proceeded to launch the first major mathematical research school in Germany.

If Jacobi's Königsberg school represented a decisive breakthrough for research-level mathematics at the German universities, it was hardly an isolated phenomenon. Indeed, Jacobi's experience attending the Berlin seminar run by the classical philologist, August Boeckh, was a major inspiration for his subsequent teaching activity in Königsberg. Moreover, Boeckh's seminar shared many common features found at other institutions, and thus should itself be seen within the broader context of educational reforms, spearheaded by progressive Prussian government officials, that transformed the German universities into prototypes of the modern research-oriented institutions found throughout the world today. Seen from a long-term view, the heart of this transformation amounted to a form of secularisation, instumentalised by *inverting* the status of the theological and philosophical faculties while freeing the members of the latter to pursue their chosen area of expertise unobstructed by the concerns of their colleagues, including the theologians. *Lehr- und Lernfreiheit*, the twin pillars of this new academic creed, gave Prussian scholars their understanding of academic freedom. For Jacobi, the pursuit of mathematics was not merely a matter of choosing a research career; like any dedicated *Wissenschaftler*, such a choice carried strong ethical overtones as a higher calling, as an endeavour through which the individual realises himself by dedicating his life to a discipline. The sense in which Jabobi and his contemporaries viewed their work as a secularised religious activity is difficult to pinpoint, of course, but sentiments of pathos can be found over and again in the speeches delivered by German professors on ceremonial occasions (a theme discussed with insight by Herbert Mehrtens in [14], a provocative study of mathematical modernity that focuses especially on fundamental tensions within the German mathematical community). No doubt such feelings were in part aroused by the hardships associated with pursuit of an academic career, including financial hardships that precluded many from even contemplating the effort.

A strong animus also came from neo-humanism, coupled with and reinforced by the flowering of German arts and letters (Goethe, Schiller, et al.). These currents intermingled with a distinctly Teutonic tradition of philosophical idealism (Fichte, Schelling, Hegel) that gave intellectual substance to the cultural expressions of Romanticism. Neohumanism, the revival of interest in classical learning, also formed the bedrock of the Prussian *Gymnasium* curriculum, wherein the rationale for mathematics instruction derived primarily from the notion that the subject promoted formal reasoning abilities and thereby supported this aspect of training in ancient languages. While grounded in the study of Greek and Latin authors, the emulation of classical ideals also served as an antidote to the materialist and utilitarian tendencies many Germans identified with the French Enlightenment tradition. Jacobi's famous criticism directed at Fourier (and, ironically, taken up by Jean Dieudonné as a personal motto in [6]), vividly captures this clash of world views: "...Monsieur Fourier was of the opinion that the principle aim of mathematics is to serve mankind and to explain natural phenomena; but a philosopher such as he ought to have known that the sole aim of science is the fulfillment of the human spirit, and that, accordingly, a question about numbers has as much significance as a question about the workings of the world" (Jacobi to Legendre, 2 July 1830, translated from [8], p. 454). Although such sentiments may sound more than a little dated today, to the ears of contemporaries words like these resonated with deeply-felt meanings.

From its founding, Berlin not only emerged as Prussia's leading university but also as one of the foremost centres of support for this neo-humanist movement. Even those whose background and interests favoured applied mathematics, like the civil engineer August Leopold Crelle, found themselves unable to resist the trend toward purism. Thus, Crelle's *Journal für die reine und angewandte Mathematik*, which he founded in 1826, was dominated from the outset by pure mathematics. The first volumes carried works by Abel, Jacobi, and Steiner that set the tone for what followed, prompting the critics' joke that Crelle ought to have called his publication *Journal für die reine, unangewandte Mathematik* (Journal for pure unapplied Mathematics, see [10], p. 95). Still, during the 1820s German mathematicians could ill afford to ignore the vital new impulses that issued from Paris, without which the sudden ascendance of higher mathematics east of the Rhine would have been unthinkable.

An especially striking example of this French influence can be seen in the early career of Gustav Peter Lejeune Dirichlet, who studied in Paris from 1822 to 1826. Working closely with Fourier and Poisson, Dirichlet also managed to catch the eye of another German expatriot, Alexander von Humboldt, whose connections with the Berlin court and Prussian ministerial officials enabled him to exert a strong influence on mathematical affairs. Knowing that the great Adrien-Marie Legendre had praised Dirichlet's work, Humboldt seized the opportunity to promote the young man's career. After a brief stint in Breslau, Dirichlet gained an appointment as *Privatdozent* in Berlin in 1829, advancing to *Extraordinarius* (associate professor) in 1831, and finally to *Ordinarius* (full professorship) in 1839. Dirichlet's arrival in the Prussian capital marks the onset of a new era in which Berlin University began to play a conspicuous part in the flowering of a distinctly German mathematical culture. Dirichlet was joined in 1834 by the Swiss geometer Jacob Steiner and then, ten years later, by the dynamic Jacobi, who became King Frederick William IV's unofficial "court *mathematicus*" through his appointment to the Prussian Academy. Together this trio of brilliant individualists stood at the

Jacob Steiner

center of a vibrant and intimate mathematical community that soon attracted impressive younger talents to Berlin. Three particularly illustrious names, connected mainly with Dirichlet during these years, were: Gotthold Eisenstein, who taught as a *Privatdozent* in Berlin until his premature death in 1852 at the age of 29; Leopold Kronecker, Berlin's most influential mathematician during the 1880s; and Bernhard Riemann, the melancholy genius whose ideas exerted such a deep impact on modern mathematics (for a well-rounded account of his life and work, see [12].)

During the events of 1848–49, the Berlin mathematicians were actively involved in efforts to promote political reform. Dirichlet, a convinced democrat, stood guard after seizure of the royal palace, while the police listed Steiner as possibly suspect in the black book that recorded information on politically subversive elements in the local population. When the crackdown came, Eisenstein found himself in the thick of the fighting and nearly lost his life from the severe beatings he suffered at the hands of Prussian soldiers. The outspoken Jacobi experienced nothing quite this traumatic, but he nevertheless received a real taste of the reactionary backlash. As a baptized Jew dependent on the largesse of the king, he found himself particularly vulnerable to royal whims. Rumors spread about a speech he had once delivered in a political club and which had caused a great stir among those present at the time. Amid reports of his political activities in the Berlin press, Jacobi scrambled to save his reputation, but his patron Frederick William decided that the matter was too serious to be ignored and suspended his

Gustav Peter Lejeune Dirichlet

salary for an indefinite period. No longer able to afford the lavish costs of housing a large family in Berlin, Jacobi made arrangements for them to live in Gotha until such time as he could regain his former standing. His salary was eventually restored, but not long afterward he fell suddenly ill from smallpox. His death in 1851 was followed by Eisenstein's a year later, and with these two sudden departures Berlin's brilliant first period began to reach its close.

The year 1855 was marked by the deaths of Gauss and Crelle, which brought about a series of changes in Berlin, beginning with Dirichlet's appointment as Gauss's successor in Göttingen. For a brief time, the future of Göttingen mathematics looked bright indeed, as both Riemann and Dedekind gravitated into Dirichlet's circle during the late 1850s (on Dedekind, see [20]). In both Berlin and Göttingen, Rebekka Mendelssohn-Bartholdy Dirichlet, sister of the famous composer, opened the mathematician's home to musicians and literati. Her presence helped the young Dedekind, a gifted pianist, to overcome his natural reservedness.

In the meantime, in Berlin the burden of filling Dirichlet's shoes fell to the master's hand-picked successor, Ernst Eduard Kummer, a former *Gymnasium* teacher who had managed to rise to a professorship in Breslau. Joining Kummer in Berlin in 1856 was another late bloomer, a forty-one-year-old *Gymnasium* teacher by the name of Karl Weierstrass who had first surfaced two years earlier after solving the Jacobi inversion problem for hyperelliptic integrals. Compared with their genial predecessors, these two former *Oberlehrer* could hardly have instilled great

Leopold Kronecker

confidence that Berlin's greatest years in mathematics lay in the years immediately ahead. Yet, supported by Kronecker, Kummer and Weierstrass soon brought about what Kurt-R. Biermann has called the "golden age" of Berlin mathematics, while Göttingen's fortunes faded after Dirichlet succumbed to a heart attack in 1859 soon after his wife's premature death. His unlucky successor, Bernhard Riemann, already contracted a lung infection in 1862 that forced him to spend most of his time thereafter convalescing in Italy. He died near Lago Maggiore in 1866 before reaching the age of forty (his close friend Dedekind had left Göttingen for Zürich back in 1858). Coincidentally, Göttingen became Prussianized in the very year of Riemann's death when the state of Hanover was annexed by the Hohenzollerns.

Meanwhile, in Berlin, Kummer and Weierstrass managed to seize a series of new initiatives. The most important of these came about in 1860 when they founded the first seminar at a German university devoted exclusively to the study of higher mathematics, a milestone event for the professionalization of mathematical research (see [1] and [2]). The success of this undertaking can be adduced from the fact that between 1862 and 1884 no fewer than 22 graduates of the Berlin seminar went on to professorships at German institutions of higher education, a figure representing a significant portion of the total number of such positions available. Indeed, during this period nearly all the vacancies at the universities in Prussia were filled by Berlin graduates, a circumstance that led to considerable embitterment among outsiders and a general polarisation of the German mathematical

Ernst Eduard Kummer

community. The rival research school established in Giessen and later Göttingen by Alfred Clebsch launched its own journal, *Mathematische Annalen*, in an effort to combat the Berlin-dominated *Crelle Journal* which was edited between 1855 and 1880 by Carl Wilhelm Borchardt. Efforts to launch a nation-wide organisation soon after the founding of the Second Empire in 1871 floundered due to the indifference of the "North German mathematicians," a euphemism for those connected with the Berlin establishment. Clebsch's premature death in 1872 left his camp in a state of disarray, but his youngest disciple, Felix Klein, quickly took steps to fill the vacuum of power. Plans for a broadly-based national organisation, however, were put on hold after a preliminary planning session held in Göttingen in 1873 failed to mobilize sufficient support.

In the wake of these turbulent affairs, Berlin emerged by the 1870s as the dominant mathematical research centre in Germany with Weierstrass as the internationally recognised leader of the era's leading school. His principal speciality was complex analysis, particularly higher transcendental functions, a field first launched with the pioneering work of Cauchy, Abel, and Jacobi, but which only emerged as a fully developed theory in the hands of the Berlin master. Weierstrass sought to build an imposing structure starting from basic properties of convergent infinite series while avoiding all use of geometrical arguments. Furthermore, he made no appeal to known results found in the published literature – everything had to be proved from scratch using only those theorems he had established along

Karl Weierstrass

the way. Since he published nearly nothing from his regularly repeated cycle of lectures, those who wished to learn his approach to complex function theory were virtually compelled to spend several semesters attending his classes in Berlin. At the height of his fame, upwards of 200 students would find their way into Weierstrass's lecture hall in hopes of grasping the mysteries of higher analysis. Few, of course, succeeded, but among those who did were a number of the next generation's leading mathematicians, including Georg Cantor, Hermann Amandus Schwarz, Leo Königsberger, Gösta Mittag-Leffler, and Sofia Kovalevskaia (on Cantor and Kovalevskaia, see [5], [15], and [3]). These circumstances, coupled with the fact that Weierstrass's work fit into a well-conceived research program far too vast to be carried out by a single individual, help account for the special school-like atmosphere that surrounded Berlin's most influential teacher.

If Weierstrass's fame had much to do with the immediate impact he exerted on the members of his school, one should not overlook Kummer's central role within the larger scope of the Berlin enterprise. A versatile mathematician, Kummer supervised the doctoral work of roughly the same number of candidates as Weierstrass during the period from 1856 to 1883 when they worked together. Nor should be forgotten the mathematical activities that went on within the Prussian Academy of Sciences during this era. Kummer served for several years as secretary of the Berlin Academy, a duty he managed with such dedication and aplomb that many came to associate his person with the office itself. On his recommendation,

Palace of Prince Heinrich

Kronecker became a full member of the Academy in 1861, which gave the latter the right to teach at the university as Jacobi had done during his final years in Berlin. Thereafter, Kronecker took an active role in Academy affairs, personally nominating fifteen prominent mathematicians as members between 1863 and 1886.

Like Jacobi, both Borchardt and Kronecker came from fairly wealthy Jewish families. The young Leopold Kronecker grew up in Silesia, where he had already known and befriended Kummer when the latter was his mathematics teacher at the *Gymnasium* in Liegnitz. Their mutual interest in higher arithmetic reinforced the striking purism typical of Berlin-style mathematics. Borchardt had studied under both Dirichlet and Jacobi before gaining an appointment as *Privatdozent* in Berlin in 1848. He was one of the privileged few, along with Sofia Kovalevskaia, whom Weierstrass addressed with the familiar *Du* form. These personal ties should be borne in mind as a prelude to the events of the 1880s when, after Borchardt's death in 1880 and Kummer's retirement in 1883, relations between Weierstrass and Kronecker began to deteriorate dramatically. To ease the transition prior to Kummer's departure, Kronecker joined the Berlin faculty as its third mathematics *Ordinarius* in 1882 while assuming the editorship of *Crelle's Journal*. Kronecker had long been overshadowed by Weierstrass, and he now finally found himself in a position from which he could hope to launch a school of his own. By this time his more famous colleague was an ailing old man who could no longer keep pace with the brilliant algebraist. For although now in his 60s, Kronecker showed no signs that he might be slowing down, and he lost few opportunities to promote his ambitious constructivist program to arithmetize algebra and analysis. Unfortunately, Weierstrass saw this missionary zeal as nothing less than an effort to undermine his own life's work in analysis. Thus, Kummer's departure was followed by a rapid disintegration of the previously congenial relations in Berlin, a state of affairs that could not bode well for the future.

Indeed, the year 1892 marks not only the end of the Kummer–Weierstrass–Kronecker era in Berlin mathematics but also the beginning of a major shift in power relations within the German mathematical community. The *Deutsche Mathematiker-Vereinigung* (DMV) had been founded two years earlier, providing a national organisation designed to serve the interests of professional mathematicians, particularly those who taught at the German universities. Kronecker took a brief, but active role in promoting the DMV in cooperation with Georg Cantor, who led the effort to organise the "North German mathematicians." Then came Kronecker's sudden, unanticipated death in late 1891, followed by Weierstrass's long-awaited retirement, thereby setting off a chain reaction of appointments in mathematics throughout the German universities. Weierstrass's chair went to his former pupil, H.A. Schwarz, whereas Kronecker's was taken by the algebraist, Georg Frobenius, another leading exponent of the Berlin tradition. Of the two, Frobenius quickly emerged as the more forceful personality, particularly since Schwarz seems to have entered a period of semi-retirement from the moment he arrived in Berlin.

Schwarz's departure from Göttingen also gave Felix Klein his first opportunity to pursue a new research orientation inspired by the two greatest names connected with the Göttingen mathematical tradition: Gauss and Riemann. With the help of the Prussian Ministry of Education and, in particular, its autocratic head of university affairs, Friedrich Althoff, Klein hoped to expand his own formidable network of contacts and somehow to use these in order to transform Göttingen into an intensely interactive scientific center (see [18]). By taking the lead in this initiative, rather than developing such a plan in cooperation with his colleagues in the natural sciences, Klein could push forward a vision of the Göttingen community in which he and his fellow mathematicians stood at the hub of the whole enterprise. One of the cornerstones in Klein's overall strategy for building up mathematics in Göttingen was reliance on diverse intellectual and material resources with the potential for mutual reinforcement. While this might sound rather trite and obvious, such a tendency ran straight across the grain of established organisational structures within the German university system. As semi-autonomous institutions, Germany's philosophical faculties, particularly in the Prussian universities, had proliferated throughout the course of the century largely by virtue of the specialised research interests typically undertaken in various scientific schools. Such schools had, of course, no official status, but they clearly took a conspicuous part within the informal scientific networks that played an increasingly prominent role in setting research agendas and influencing careers. Mathematics, like many other disciplines practiced at the German universities, largely depended on the local conditions that prevailed at these institutions. Academic appointments often followed established patterns of interlocking relationships, many of which had their roots in school affiliations. Those mathematicians identified with the Berlin tradition in the era of Kummer and Weierstrass thus enjoyed the potential advantages of affiliation with a strong, established network that stretched well beyond the borders of Germany.

Klein, who instinctively understood the importance of domestic and foreign power relations in mathematics, always had great respect for this Berlin tradition. By the same token, he had little use for the kinds of narrowly focused research schools that adorned many German universities; nor did he advocate the kind of disciplinary purism promoted in Berlin, sensing that this tendency weakened the position of mathematics vis-à-vis other neighboring disciplines. His Göttingen experiment thus meant consciously breaking with the notion of "mathematics for mathematics sake" which had served as the dominant ideology of German mathematicians since Jacobi's day. Nevertheless, Klein had no interest in undermining traditional pure mathematics; he thought of Göttingen mathematics as representing in microcosm the full range of the discipline, including number theory, algebra, and higher analysis, thereby paying Gaussian rigour its rightful due. Above all, he was concerned that the nearly exclusive interest in these traditional pillars of strength had led to an unhealthy imbalance and, in particular, the neglect of applications. In Klein's view, it was high time for the German mathematical community to climb down from its ivory tower and pay more attention to the needs of modern exact scientists and engineers. Frobenius, who saw himself as the prime protector of the Berlin legacy, loathed everything associated with Klein and his Göttingen enterprise. He and the other leading representatives of Berlin mathematics – Schwarz, Fuchs, Schottky, and Schur – thus took no part in Klein's massive *Encyklopädie der mathematischen Wissenschaften*, a project Frobenius once dubbed as "senile science" (F. Klein to Walther von Dyck, 24 March 1893, Bayerische Staatsbibliothek, München, cited in [18], p. 209). As it turned out, though, he was fighting a hopeless uphill battle against a far shrewder man. Klein took a decisive step toward strengthening pure mathematics in Göttingen in late 1894 when he arranged David Hilbert's appointment as successor to Heinrich Weber. By this time, Klein had also won Althoff's confidence and backing for his far-reaching plans to tap private funding sources, in particular from German industry, in order to build scientific research institutes in Göttingen. Hilbert's vast research program, unveiled in part in his famous address of 1900 at the Second International Congress of Mathematicians in Paris, fit beautifully within Klein's overall conception. If contemporary mathematicians continued to identify the massive talent that gathered around Hilbert after 1900 as the "Hilbert school," we can now with retrospect recognize the striking differences between this phenomenon and conventional mathematical schools. Indeed, Hilbert openly criticized the kinds of schools that dominated the German universities during the 1880s and 1890s, advocating a more integrated model for organising mathematical research.

This latent power struggle between Göttingen and Berlin came to a head in 1902 when Frobenius and Schwarz nominated Hilbert for the chair vacated by the death of Lazarus Fuchs. Not since Gauss, had a German mathematician turned down a call to Berlin, and Hilbert appears to have given the matter serious consideration. But Althoff quickly ended such speculation by creating a third professorship in Göttingen to be occupied by Hilbert's close friend, Hermann Minkowski, a move Hilbert later effusively called "unique in the annals of Prussian higher education" ([7], p. 355). Fuchs's chair went to Friedrich Schottky, a personal friend of Frobenius whose best years as a mathematician were well behind him. Klein and Hilbert thereafter moved quickly to consolidate their advantage. Two years later, Carl Runge and Ludwig Prandtl joined the Göttingen faculty, thereby ensuring its dominant standing in applied as well as pure mathematics. Even the Prussian Ministry was apparently taken by surprise. One of Althoff's colleagues wrote to Klein afterward: "After Minkowski's appointment, Hilbert said 'now we are invincible'. But you continue right on becoming ever more invincible, so that I am really curious what proposals you will come up with next. But we are prepared for anything as nothing can surprise us anymore" (Elster to Klein, 18 July 1904, cited in [18], p. 197). Althoff's policy of creating centers of disciplinary excellence together with Klein's broad conception of what a modern mathematical research center might be, thus helped create the preconditions in Göttingen for an unprecendented explosion of mathematical activity that soon made Berlin look like a provincial university by comparison.

Minkowski's sudden death in 1909 came as a shock, but the net result was yet another loss for Berlin when Edmund Landau was appointed his successor. Landau had been a *Privatdozent* in Berlin since 1901 and, along with Issai Schur, he embodied the purist ethos that had long been the hallmark of the Berlin tradition. Closer to Schwarz than to Frobenius, Landau continued to maintain ties with his former colleagues in Berlin, hoping that he would eventually be called back to a full professorship. Soon after the end of the war, however, these hopes faded, forcing Landau, who identified heart and soul with the Berlin mathematical tradition, to reconcile himself with his fate in Göttingen. Naturally, the "war to end all wars" radically altered mathematical life in both Göttingen and Berlin. As the casualty reports came in, Einstein's general theory of relativity served to galvanise the mathematical community, spawning efforts by Hilbert, Klein, Emmy Noether, and numerous others. Once again, the Göttingen mathematicians left their Berlin counterparts standing on the sidelines, despite the fact that Einstein himself was an active member of the Berlin Academy! As another sign of the times, the death of Frobenius in 1917 received not even so much as a small obituary notice in any of the leading publications of the German mathematical community, which routinely published such articles for members of far less distinction and accomplishment.

Soon afterward Erhard Schmidt and Constantin Carathéodory, two close protégés of Hilbert, assumed the chairs formerly occupied by Schwarz and Frobenius, respectively. Carathéodory had been Klein's successor in Göttingen before coming to Berlin in 1918; he remained, however, for only one year. Following his departure, the Berlin mathematicians faced the same difficulty their Göttingen counterparts experienced that very year when Erich Hecke left for Hamburg, namely, the problem of filling prestigious chairs in Germany during the turmoil of the immediate post-war period. As it turned out, the preferred candidates to replace both Carathéodory and Hecke were virtually identical: both Schmidt and

Erhard Schmidt

his former mentor, Hilbert, hoped to gain the services of the Dutch topologist, L.E.J. Brouwer (the outbreak of the *Grundlagenstreit* was yet to come). Brouwer thus stood at the top of the Berlin *Berufungsliste*, followed by Zürich's Hermann Weyl, and Gustav Herglotz, then stationed in Leipzig. Just as in Göttingen, where Richard Courant was eventually chosen after Brouwer and Weyl declined, all three candidates refused the Berlin chair, as did Hecke afterward. This finally opened the way for the Frankfurt function-theorist, Ludwig Bieberbach, who accepted the call in 1921. In that same year, Schottky also retired, enabling Schur, who had been *Extraordinarius* since 1916, to attain a full professorship. Alongside the three professorships in pure mathematics came a fourth applied position, occupied in the meantime by the talented probabilist Richard von Mises. This appointment, not surprisingly, was engineered by the former Göttingen star, Erhard Schmidt, who thereby achieved a major breakthrough for applied mathematics in Berlin.

During the ensuing Weimar years, Berlin regained much of the prestige it had lost to Göttingen during the heyday of the Klein–Hilbert era. Together Berlin's four *Ordinarien* – Schmidt, Schur, Bieberbach, and von Mises – provided quality, breadth, and balance. Their presence also drew a number of gifted younger mathematicians, thereby breaking Göttingen's long-standing monopoly as a training ground for the next generation's brightest talents. Among the better-known figures who settled into this new Berlin milieu were the *Privatdozenten*: Gabor Szegö (1921–26), Karl Löwner (1923–28), Heinz Hopf (1926–31), John von Neumann (1927–32), Robert Remak (1929–33), Eberhard Hopf (1929–32), Stephan Bergmann (1932–33), and Alfred Brauer (1932–35). As personalities, the regal Schmidt as well as the unassuming Schur contrasted sharply with the more temperamental Bieberbach, but whatever tensions may have lurked beneath the surface, outwardly the Berliners projected a harmonious image of their mutual relations. Bieberbach's *bête noire* from as early as 1920 was Göttingen's Edmund Landau, who after Frobenius's death cast himself in the awkward role – given his physical dislocation – of the era's most outspoken advocate of Berlin's purist ideals. Equally ironic, the Berliner Bieberbach saw himself as upholding the tradition of *anschauliche Mathematik* as championed earlier by Felix Klein, a position he, Bieberbach, later exploited as a turncoat Nazi to Landau's detriment (see [13]). Bieberbach's animus toward Landau can already be detected in his response to a remark made by the latter at the 1920 DMV meeting held in Bad Nauheim, where Landau was elected president of the organisation. Possibly in the discussion following Bieberbach's lecture concerning recent problems in the theory of conformal mappings, Landau offered a definition of a geometer as "a man who regards false theorems as self-evident." Countering this *geistreiche Definition* in a review published in the same year, Bieberbach appealed to the famous words of Clebsch in describing Plücker's geometrical genius: "*es ist die Freude an der Gestalt in höherem Sinne, welche den Geometer macht*" (Ludwig Bieberbach, review of H. Beck, Koordinatengeometrie, Jahresbericht der Deutschen Mathematiker-Vereinigung 29 (1920), 63; see the discussions by Mehrtens in [13], [14]). In general, however, relations between Germany's two leading mathematical centers improved noticably during the early 1920s. During these years, Göttingen's Richard Courant emerged as the era's new entrepreneur and Ferdinand Springer as its new publishing mogul, both finding in Erhard Schmidt and his Berlin colleagues generally co-operative partners in matters of mutual concern and interest.

This brief period of tranquillity ended rather abruptly, however, following a series of dramatic events in 1928, beginning with the publication of a letter from Brouwer to the DMV in which he urged its members to boycott the forthcoming International Congress of Mathematicians in Bologna. The two previous ICMs – held in Strasbourg (1920) and Toronto (1924) – had taken place under the auspices of the *Union Mathématique Internationale*, which forbade German participation. Brouwer argued that the organisers of the Bologna Congress had failed to make a clean break with the UMI's political line, thereby tainting the ideal of a truly international community of mathematicians. Bieberbach quickly threw his support behind Brouwer's call for a German boycott, but he was immediately countered by Hilbert, then seriously ill with pernicious anemia. Despite his precarious health, Hilbert was determined to thwart the boycott effort, and so travelled to Bologna where in a rousing speech he declared: "It is a complete misunderstanding of our science to construct differences according to peoples and races ... Mathematics knows no races ... for mathematics, the whole cultural world is a single country" (quoted in [16], p. 188). In view of Bieberbach's later activities, one cannot read these words without wondering whether Hilbert already sensed the direction his Berlin rival might have been heading.

Immediately after this event, Hilbert unilaterally purged Brouwer from the editorial board of *Mathematische Annalen*, despite the well-intended advice of nearly all his closest associates (for details, see [4]). When the dust had cleared, Bieberbach's name, too, no longer appeared on the title page of the world's most prestigious mathematics journal. This traumatic episode not only severed Brouwer's ties with Göttingen it also served to repolarise the German mathematical community, this time along largely political lines. The atmosphere in the Göttingen mathematical community reflected many typical elements of the Weimar era's counter-culture with its intermingling of Jews, foreigners, and even some women (on Göttingen mathematics as a Weimar culture phenomenon, see [17]). Similar tendencies pervaded Berlin's *Mathematisch-physikalische Arbeitsgemeinschaft* (*Mapha*), a students' organisation headed by Alfred Brauer and Hans Rohrbach with a relatively high percentage of Jewish members. But within the Berlin faculty nothing comparable to Göttingen's open-ended, internationalist tendencies surfaced, as reflected by the fact that both Schmidt and von Mises supported Bieberbach's call for a boycott of the Bologna Congress. After this effort backfired, Bieberbach engaged in another losing power struggle with the Göttingen mathematicians over control of review journals. This began when he engaged the Prussian Academy to undertake a series of measures to bolster the Berlin-based *Jahrbuch über die Fortschritte der Mathematik*, originally founded in 1869. Bieberbach then reacted in dismay when Courant began making plans with Springer in 1930 to launch a new review journal, the *Zentralblatt* (see [21]). This latter undertaking had barely gotten off the ground, however, when both Courant and the *Zentralblatt's* main workhorse, Otto Neugebauer, were compelled to flee Nazi Germany.

The political collapse of Weimar deeply affected the careers of innumerable mathematicians, particularly those at the bottom of the academic ladder. In Berlin, Remak, Bergmann, and Brauer were all forced to give up their positions after the Nazi government came to power; Löwner, and Bergmann would join Szegö much later at Stanford (see [19]). Among the senior faculty, Isaai Schur, Berlin's last representative of an algebraic tradition reaching back to Kummer, Kronecker, and Frobenius, was forced to enter early retirement, emigrating to Palestine in 1939. Politics had once again come to shape the course of mathematical events in Berlin, but this time, after the brief renaissance of the Weimar years, the effects were anything but positive and Berlin soon lost its once formidable place in the cultural world of mathematics.

Acknowledgements

The standard study upon which much of this essay draws is Kurt-R. Biermann's [1]. The author takes pleasure in thanking Moritz Epple, Helmut Koch, Norbert Schappacher, and Reinhard Siegmund-Schultze for their insightful comments on an earlier version of this essay.

References

[1] *Biermann, K.-R.*, Die Mathematik und ihre Dozenten an der Berliner Universität, 1810–1933, Akademie Verlag, Berlin, 1988.

[2] *Calinger, R.*, The Mathematics Seminar at the University of Berlin, in: Calinger, R. (ed.), Vita Mathematica, The Mathematical Association of America, Washington, 1996, 153–176.

[3] *Cooke, R.*, The Mathematics of Sonya Kovalevskaya, Springer-Verlag, New York, 1984.

[4] *van Dalen, D.*, The War of the Frogs and the Mice, or the Crisis of the Mathematische Annalen, The Mathematical Intelligencer **12** (4) (1990), 17–31.

[5] *Dauben, J.W.*, Georg Cantor, His Mathematics and Philosophy of the Infinite, Harvard University Press, Cambridge, Mass., 1979.

[6] *Dieudonné, J.*, Pour l'honneur de l'esprit humain, Hachette, Paris, 1987; Engl. trans., Mathematics – The Music of Reason, Springer-Verlag, Heidelberg, 1992.

[7] *Hilbert, D.*, Gedenkrede auf Hermann Minkowski, Math. Ann. **68** (1910), 445–471, reprinted in *Hilbert, D.*, Gesammelte Abhandlungen, 3 vols., Springer-Verlag, Berlin, 1932, 1933, 1935, vol. 3, 339–364.

[8] *Jacobi, C.G.J.*, Gesammelte Werke, vol. 1, Georg Reimer, Berlin, 1884.

[9] *Kaufmann-Bühler, W.*, Gauss, A Biographical Study, Springer-Verlag, New York, 1981.

[10] *Klein, F.*, Vorlesungen über die Entwicklung der Mathematik im 19. Jahrhundert, 2 vols., Springer-Verlag, Berlin, 1926–1927.

[11] *Koenigsberger, L.*, C.G.J. Jacobi, Teubner, Leipzig und Berlin, 1904.

[12] *Laugwitz, D.*, Bernhard Riemann, 1826–1866, in: Vita Mathematica, vol. 10, Birkhäuser, Basel, 1996.

[13] *Mehrtens, H.*, Ludwig Bieberbach and Deutsche Mathematik, in: Phillips, E. (ed.), Studies in the History of Mathematics, The Mathematical Association of America, Washington, 1987, 195–241.

[14] *Mehrtens, H.*, Moderne-Sprache-Mathematik, Suhrkamp, Frankfurt, 1990.

[15] *Purkert, W., Ilgauds, H.J.*, Georg Cantor, 1845–1918, in: Vita Mathematica, vol. 1, Birkhäuser, Basel, 1987.

[16] *Reid, C.*, Hilbert, Springer-Verlag, New York, 1970.

[17] *Rowe, D.E.*, Jewish Mathematics at Göttingen in the Era of Felix Klein, Isis **77** (1986), 422–449.

[18] *Rowe, D.E.*, Klein, Hilbert, and the Göttingen Mathematical Tradition, in: Olesko, K.M. (ed.), Science in Germany: The Intersection of Institutional and Intellectual Issues, Osiris **5** (1989), 189–213.

[19] *Royden, H.*, The History of the Mathematics Department at Stanford, in: Duren, P. (ed.), A Century of Mathematics in America, Part II, American Mathematical Society, Providence, R.I., 1989, 237–278.

[20] *Scharlau, W. (ed.)*, Richard Dedekind, 1831–1891, Vieweg, Braunschweig-Wiesbaden, 1981.

[21] *Siegmund-Schultze, R.*, Mathematische Berichterstattung in Hitlerdeutschland, Vandenhoeck & Ruprecht, Göttingen, 1993.

Augustus Leopold Crelle, 1780–1855

Jean-Pierre Friedelmeyer

Introduction

All mathematicians know and read *Crelle's Journal*. Few will know that its founder, Augustus Leopold Crelle, was one of the most eminent personalities in Berlin society of his time.

Crelle was born to a dike reeve on March 11th, 1780, in Eichwerder near Wriezen an der Oder. He was mainly self-taught and worked as a civil engineer, but his qualities were such that he was rapidly appointed as public buildings chief engineer and member of the Prussian Administration Headquarters. Crelle developed an interest in mathematics very early on, and published a number of works on a great variety of subjects, mainly applied and school mathematics and science. He translated Legendre's *Géométrie* and Lagrange's works from French into German.

His lifetime's work was created in 1826: the *Journal für die reine und angewandte Mathematik*, better known to this very day as *Crelle's Journal* (even though the 19th century habit of calling journals by their chief editor has otherwise been lost). Although *Crelle's Journal* was not the first journal exclusively devoted to mathematics [4], it initiated and promoted a new way to circulate scientific information. Instead of relying on the Academies' publications (like the *Transactions of the Royal Society*, the *Mémoires de l'Académie des Sciences de Berlin, de Paris*, or *de St. Petersbourg*, etc.), the circulation of scientific discoveries and inventions was now going to be ensured mostly through the publication of periodicals. This evolution reflects the growing professionalisation of the scientists' world. The creation of the *Grandes Ecoles* (*Polytechnique, Normale Supérieure, Ecoles Centrales*) in France, and W. Humboldt's neo-humanist reform in Prussia, led to the appointment of professors who were to devote their lives to pure scientific research and teaching.

At the University of Berlin, several factors combined to lead to the birth of an important Mathematics Research Center in the late 1820s at the decisive prompting of two exceptional personalities: Alexander von Humboldt and Augustus Leopold Crelle. Humboldt's instrumental role can be seen in other articles of this volume. His idea was to create in Berlin a major science center with "the first observatory, the first chemistry laboratory, the first botanical garden, the first school of transcendental mathematics" [1]. The promotion of mathematics was to a very large extent Crelle's doing, through his journal.

Benefiting simultaneously from the profound philosophical pondering initiated by Kant over the conditions and modalities of a pure science, from the movement raised by the German romanticism and neo-humanism and from the setting of a city in full cultural and scientific boom, the exemplary success of

August Leopold Crelle

Crelle's Journal can be understood as the happy encounter between the *Zeitgeist* of the period with the remarkable personality of its founder who continued to energetically control his creation, and who continued to be active in official institutions.

A very rich human personality, attractive and both self-asserted and adaptable

All those who met Crelle were impressed by his extreme friendliness. Let us simply quote Abel's appreciation in a letter to Holmboe (January 1826): "You cannot imagine what an excellent man he is, exactly as one should be, thoughtful and yet not horribly polite like so many people, quite honest, for that matter. I am with him on as good terms as I am with you or other very good friends." More to the point for the success of his journal, Crelle had an extraordinary intuition for judging the qualities of young talents and for encouraging them with their research work. Few publishers can claim to have their name associated with the glory of as many first-rate young mathematicians as Crelle: the first works of Abel, Dirichlet, Eisenstein, Grassmann, Hesse, Jacobi, Kummer, Lobachevski, Moebius, Plücker, von Staudt, Steiner, and Weierstrass were published in *Crelle's Journal*.

But his sociable personality did not prevent him from being strongly success-oriented. He invested an enormous amount of work, personal money, indeed even his health into the journal. The huge acclaim which the first issue received, rewarded him and certainly encouraged him to continue. The success was immedi-

Alexander von Humboldt

ately international, as A. von Humboldt reports after one of his trips to Paris: "*Crelle's Journal* is much highly praised in France where it is preferred to Mr. Gergonne's journal. I learnt that while I was in France: it is constantly quoted at the Institute (i.e., the Academy of Science) where MM. Laplace, Fourier, Cauchy, Poinsot, Le Gendre publicly gave proof of their high esteem" [3].

A personality involved in the social and cultural life of his time

Abel reports in a letter to Hansteen that Crelle held salons every Monday evening: "There is at his place some kind of meeting where music is mainly discussed, of which unfortunately I do not understand much. I enjoy it all the same since I always meet there some young mathematicians with whom to talk. At Crelle's house, there used to be a weekly meeting of mathematicians, but he had to suspend it because of a certain Ohm (Martin, a brother of Georg, the physicist) with whom nobody could get along due to his terrible arrogance."

Crelle had extensive international contacts, in particular in France. In 1830 he travelled to Paris. He met Gauss in Göttingen several times and had him write some articles for his journal. He was not only a member of the Academy of Berlin, but also a corresponding member of the Science Academies of Saint-Petersburg,

Naples, Brussels, Stockholm, of the American Philosophical Society and the Mathematical Society of Hamburg.

An actor in the official institutions

Indeed, Crelle was in the first place an influential civil servant in the field of building and public works and he played a major role in the railroad construction in Germany in the 1830s. As such he created also another journal, the *Journal für die Baukunst*, whose publication in thirty volumes he ensured between 1829 and 1851. In 1829 he was appointed Private Counsellor to the Ministry of Education. His activities then took two directions: He elaborated a scheme for the creation of a polytechnical institute for mathematics, physics and chemistry that followed the pattern of the *Ecole Polytechnique Française* which greatly fascinated the German intellectuals; its purpose was to train professors. The project was not realized. He perfected a curriculum for the teaching of mathematics in Prussian high schools. In fact, he served as a chairman for school textbooks over twenty years [5].

The creation of the Journal für reine und angewandte Mathematik and its success

In the preface to the first issue, Crelle states his goals in detail. Besides the professionalisation of the scientific community mentioned above in this article, his purposes reveal another important evolution: unlike the publications of the Academies, with no direct financial worry, the existence and regularity of publication of a journal are subject to heavy economic constraints and mostly depend on the matching of its contents to the preoccupations and issues of its time. For *Crelle's Journal* it is no longer essential to allow scientists to inform one another but rather to form a group of readers who would represent science and its development. That is why the journal "must endeavour to offer itself to a larger public so as to first and foremost ensure its longevity and the possibility to perfect itself" [2]. Then, the journal cannot be devoted only to connoisseurs or merely attempt to influence the development of science; it must also concern itself with its circulation. It must be independent from trends, schools, habits, so as to open the way to all new ideas; it must be open to foreign publications which will be translated; it will deal with pure mathematics as well as applied mathematics; it will circulate the translations or reprints of major texts.

As the scientists' preoccupations are in constant flux, the continuation of *Crelle's Journal* up to today poses a fundamental question for the historian: how can such a sustained success be explained? Indeed, the success went well beyond Crelle's expectations, showing how much his journal was an answer to an essential need of the science of his time. The success was threefold: in terms of the readership and the support he got among the scientists; in the moral and sometimes financial support from the authorities; and in the quantity and quality of the proposed articles which rolled in from all over the world.

Crelle ran the journal himself until 1855. During that period of time, 52 issues were published, with 24 % of the articles written in French, 12.1 % in Latin, and 1.1 % in English or Italian.

Unlike Gergonne, whose *Annales de Mathématiques pures et appliquées* presented the articles under a multitude of untidy headings, Crelle adapted to the profound mutations of 19th century science.

The one account on which he failed in his editorial policy was that he could not maintain the balance between pure and applied mathematics he had wished for. The very evolution of the science forbade it and the theoretical importance of the articles dealing with pure mathematics soon forced him to opt for specialisation. Crelle was intelligent enough not to oppose that evolution, turning to the *Journal für die Baukunst* for the practical and technical questions. That change allowed *Crelle's Journal* to have the priority to publish quite a number of major 19th century texts:

- *Recherches sur la série* $1 + \frac{m}{1}x + \frac{m(m-1)}{1\cdot 2}x^2 + \frac{m(m-1)(m-2)}{1\cdot 2\cdot 3}x^3 + \dots$ (Abel)
- *Elliptic functions* (Abel and Jacobi)
- *Sur la résolution des équations dont le degré dépasse cinq* (Abel)
- *Geometry* (Moebius, Poncelet, Steiner, Plücker, Grassmann, ...)
- *Non-Euclidean geometry* (Lobachevski)
- *Number theory* (Dirichlet, Eisenstein, Kummer, ...)

It was an evolution which revealed the quasi-perfect sensitivity of *Crelle's Journal* to the prevailing ideas of the philosophical scientific and romantic thought of 19th century Germany.

The relation between pure and applied mathematics was evolving as well. From ancient times until Descartes, geometric space was an inert space, totally abstract and lifeless. Little by little, the 19th century endowed it with life, with dimensions, directions, tensions, giving birth to an ever expanding space (think of Grassmann's title *Ausdehnungslehre*) a space loaded with electric and gravitational forces upon which the numbers themselves become active under the form of operators (vectors, complex numbers, quaternions). It was the effort of abstraction undertaken by mathematicians which conditioned its increasingly performant applicability through mathematical physics. Finally, that new relation of science to objective reality made the encounter and the reciprocal influence of scientific thought and artistic production possible, in so far as both science and art edged away from the common and concrete experience, in so far as the two sublimate this common experience into shapes that are no longer mainly conditioned by this experience but by such ideal (aesthetic or scientific) points of view as simplicity, harmony, symmetry, structure. This encounter of science and aesthetics is perfectly illustrated, for example, by the emblematic figure of German romanticism, F.V.

Hardenberg, known as Novalis who will exalt the mathematical genius in so far as it is akin to the poetic genius "in its capability of treating imaginary objects as real."

Conclusion

Crelle's Journal is the most revealing mirror of the significant mutations of 19th century mathematics. Like the other journals, it shows the professionalisation of the mathematician's occupation. Unlike them, it records the break with the recourse to geometrical intuition and confirms the concept of pure mathematics as being a totally a priori knowledge. Borne by the idealist stream of German romanticism, Crelle defended a conception of mathematics which found the source of its development in itself. From his journey to Paris in 1830, he left some travel notes wherein one reads, in particular: "The real purpose of mathematics is to be the means to illuminate reason and to exercise spiritual forces" [6].

References

[1] *Biermann, K.-R.*, Die Mathematik und ihre Dozenten an der Berliner Universität 1810–1933, Akademie Verlag, Berlin, 1988.

[2] *Crelle, A.L.*, Vorrede des ersten Bandes, Journal für die reine und angewandte Mathematik, 1826.

[3] *Eccarius, W.*, Augustus Leopold Crelle als Herausgeber des Crelleschen Journals, J. Reine Angew. Math. **286–287** (1976), 5–25.

[4] *Friedelmeyer, J-P.*, La création des premières revues scientifiques, L'Ouvert, Journal de l'APMEP d'Alsace et de l'Irem de Strasbourg **86** (1997), 9–27.

[5] *Jahnke, H.N.*, Mathematik und Bildung in der Humboldtschen Reform, Dissertation Univ. Bielefeld, Institut für Didaktik der Mathematik, 1989.

[6] *Lorey, W.*, Augustus Leopold Crelle zum Gedächtnis, J. Reine Angew. Math., **157** (1927), 3–11.

Gustav Peter Lejeune Dirichlet

Helmut Koch

Dirichlet was born on February 13, 1805, in Düren, a town midway between Aachen und Cologne, where his father was town postmaster. His grandfather was a textile manufacturer from the nearby French-speaking Belgian town Richelet, so that the origin of the family name Lejeune Dirichlet is easily explained as "Young from Richelet". His parents provided him with a good education, sending him in 1817 to the *Gymnasium* in Bonn and two years later to the Jesuit College in Cologne. At the age of sixteen he completed the *Abitur*. Already during his schooling Dirichlet had developed a passion for mathematics that determined him to study this domain at the university. But since, at that time, the level of teaching at German universities was low, he went to Paris, then the centre of mathematics with such eminent scientists as Fourier, Lacroix, Laplace, Legendre and Poisson. Dirichlet came in closer contact with some of them, in particular his contacts with Fourier became important for his later research. But even more important was his study of Gauss's *Disquisitiones arithmeticae.* He was the first to understand this fundamental work in number theory, which he simplified in his later lectures in Berlin, subsequently published by Dedekind.

Beginning in the summer of 1823 Dirichlet lived in the house of General M.S. Foy (1775–1825), who was the leader of the liberal opposition in the Chamber of Deputies. Dirichlet gave German lessons to the wife and children of Foy.

In 1825 Dirichlet presented his first scientific paper *Sur l'impossibilité de quelques équations indeterminées du cinquième degré* to the French Academy of Sciences. Legendre, one of the referees, was able to prove the Last Theorem of Fermat for exponent 5 on the basis of this paper. By means of this result, Dirichlet became known in the scientific circles of Paris. He came in closer contact with Alexander von Humboldt, who lived at the time in Paris. On the recommendation of Humboldt, Dirichlet obtained the position of a *Privatdozent* in 1827 at the University of Breslau with an annual income of 400 *Thaler*. But the level of mathematics at this newly founded university was low. Therefore, Dirichlet was happy to switch to the *Allgemeine Kriegsschule* in Berlin in 1828 with the permission to give lectures at the University of Berlin. In 1829, his famous paper about Fourier series appeared, which we will come back to later in this chapter. In the same year his annual income was raised to 600 *Thaler*, and this was sufficient to establish a family in 1832. Dirichlet married Rebecka Mendelssohn-Bartholdy. A.v. Humboldt had introduced him into the house of her father, the banker Abraham Mendelssohn-Bartholdy, whose father was the philosopher Moses Mendelssohn and whose son was the celebrated composer Felix Mendelssohn-Bartholdy.

In the same year, 1832, Dirichlet became a member of the Prussian Academy of Sciences. Between 1833 and 1855 he gave a talk every year in the Academy about his own results, with the exception of the time between the autumn of

Gustav Peter Lejeune Dirichlet

1843 and the spring of 1845, when he was permitted to take a leave of absence to visit Italy. He moved with his family to Rome, where he lived together with other mathematicians from Berlin: C.G.J. Jacobi, who went to Italy to improve his health, C.W. Borchardt and J. Steiner. Steiner was accompanied by the Swiss teacher L. Schläfli, who later became a well-known mathematician. Schläfli received daily mathematical lectures from Dirichlet during this time.

In 1855 the University of Göttingen offered Dirichlet the chair of Gauss, who had died that same year. At the time of this offer, besides his lectures as professor at the university in Berlin, Dirichlet still had to give lectures in the *Kriegsschule*. Therefore, he used this occasion to ask the Prussian Ministry of Culture whether he could be freed from these classes. Since the ministry did not act in time, he accepted the offer from Göttingen and moved in the fall of 1855 to that town in the Kingdom of Hanover. He enjoyed the quiet life in this small town, but not for long. In the summer of 1858 he travelled to a meeting in Montreux, Switzerland, to deliver a memorial talk on Gauss. While there, he suffered from a heart attack and was barely able to return to Göttingen. During his illness, his wife died of a stroke, and he himself died the following spring.

Dirichlet spent twenty-seven years of his active scientific life in Berlin and exerted a strong influence on the development of mathematics through his lectures, his many pupils, the most famous of whom in Berlin were Eisenstein (1823–1852) and Kronecker (1823–1891), and through his publications, the most important

Rebecca Dirichlet

having all been written in Berlin. He did not publish very much, but all his results and proofs are of such perfection that they have been seldom simplified or improved upon.

In the following I will say something about three of his results, which may be considered as the most famous of his mathematical achievements.

1. In 1807, J. Fourier gave a lecture in the French Academy of Sciences about heat transport in solid bodies. He published a book in 1822, where he explained his ideas in more detail. In section VI of this book, he presents an arbitrary odd function $\varphi(x)$ defined on the interval $[-\frac{\pi}{2}, +\frac{\pi}{2}]$ in the form

$$\varphi(x) = \sum_{\nu=1}^{\infty} a_\nu \sin \nu x \tag{1}$$

and asserts that the coefficients a_ν are given in the form

$$a_\nu = \frac{2}{\pi} \int_{-\pi/2}^{+\pi/2} \varphi(x) \sin \nu x \mathrm{d}x. \tag{2}$$

Dirichlet published a paper in 1829 which gives an exact description of functions $\varphi(x)$ for which the integrals (2) exist and gives the coefficients of the development (1) of $\varphi(x)$ as a "Fourier series". More precisely, he proves the following:

Gustav Peter Lejeune Dirichlet

Theorem 1. *Let $f(x)$ be a piecewise continuous and monotone function $f(x)$ in the open interval $(-\pi, \pi)$. Then the series*

$$\frac{a_0}{2} + \sum_{\nu=1}^{n} (a_\nu \cos \nu x + b_\nu \sin \nu x)$$

with

$$a_\nu = \frac{1}{\pi} \int_{-\pi}^{\pi} f(x) \sin \nu x \mathrm{d}x, \quad b_\nu = \frac{1}{\pi} \int_{-\pi}^{\pi} f(x) \cos \nu x \mathrm{d}x$$

converges to the function

$$\frac{1}{2} \left(f(x+0) + f(x-0) \right),$$

where

$$f(x+0) := \lim_{\substack{y \to x \\ y > x}} f(y)$$

and

$$f(x-0) := \lim_{\substack{y \to x \\ y < x}} f(y).$$

B. Riemann wrote his *Habilitation* about Fourier series in 1854, which contained, among other things, for the first time an exact general definition of the integral now called the Riemann integral. In his introduction he wrote the following about Dirichlet's results: *Bei ihrer Zusammenstellung war es mir vergönnt, einige Winke des berühmten Mathematikers zu benutzen, welchem man die erste gründliche Arbeit über diesen Gegenstand verdankt.* (Composing this thesis I could use some hints of the famous mathematician, who wrote the first profound paper about the subject.)

Concerning the problem of presenting an arbitrary function by means of sine and cosine functions he continued: *In der That für alle Fälle der Natur, um welche es sich allein handelte, war sie vollkommen erledigt, denn so groß auch unsere Unwissenheit darüber ist, wie sich die Kräfte und Zustände der Materie nach Ort und Zeit im Unendlichkleinen ändern, so können wir doch sicher annehmen, daß die Funktionen, auf welche sich die Dirichletsche Untersuchung nicht erstreckt, in der Natur nicht vorkommen.* (In fact, in all cases of nature the problem was completely settled. As comprehensive as our ignorance about alteration of forces and states of matter in place and time in the infinitesimal may be, we can assume for sure that the functions, not included in the research of Dirichlet, do not exist in nature.)

Riemann justified his study of functions more general than those studied by Dirichlet by pointing out their importance for the foundation of analysis and number theory. He proved that an integrable function has a Fourier series development if it has only finitely many maxima and minima.

Since that time not only has there been an enormous increase in the amount of literature about Fourier series, but the ideas of Fourier, Dirichlet, Riemann and others form a basis for the development of several theories as the theory of real functions, harmonic analysis and representation theory of locally compact groups, and functional analysis.

2. Dirichlet's favourite subject was number theory. Though Euler was the first to apply methods of analysis and in particular function theory to number theory, one can consider Dirichlet as the founder of analytic number theory. Perhaps his most famous result in this direction is the following theorem about primes in arithmetic progressions published in 1837.

Theorem 2. *Let k and l be relatively prime natural numbers. Then there are infinitely many primes p such that $p \equiv l (\mathrm{mod}\, k)$.*

This theorem had long been conjectured, but before Dirichlet it was proved only for a few numbers k, such as $k = 4$. Legendre used the conjecture in his "proof" of the quadratic reciprocity law, first proved unconditionally by Gauss in 1801.

The application of analysis in Theorem 2 consists in the consideration of the series

$$L(s, \chi) := \sum_{n=1}^{\infty} \frac{\chi(n)}{n^s} \tag{3}$$

where s is a real number and χ a Dirichlet character, i.e., a multiplicative function whose values depend only on the residue class of n mod a such that $\chi(n) \neq 0$ if n is prime to a and $\chi(n) = 0$ if n and a have a common divisor. The case $a = 1$ was known to Euler, who used the series to give a new proof that there are infinitely many primes. Dirichlet followed the ideas of Euler, but he needed much deeper considerations. As in the case of Euler, the series (2) is convergent for $s > 1$, in which case it can be developed into a product (today called the Euler product):

$$L(s, \chi) = \prod_p \left(1 - \frac{\chi(p)}{p^s}\right)^{-1},$$

where the product runs over all primes p. It is rather easy to show that $\lim_{s \to 1} L(s, \chi)$ exists and is a finite number, if χ is not trivial. But the essential fact which one needs for the proof of Theorem 2 is the fact that $L(1, \chi) \neq 0$. One method of Dirichlet to prove this in the most difficult case of a real character χ consists in showing that $L(1, \chi)$ appears as a factor in a formula for the class number of quadratic forms.

The methods of analytic number theory introduced by Dirichlet play an immense role in modern number theory and analysis. Series $\sum_{n=1}^{\infty} \frac{a_n}{n^s}$ with complex coefficients a_n are called Dirichlet series. Functions generalizing $L(s, \chi)$ are now associated to a variety of mathematical objects and are the subject of important theorems and conjectures.

3. Dirichlet was the first mathematician who systematically studied rings of algebraic numbers with the goal to generalize the unique decomposition of natural numbers as products of primes.

Assuming that α is an integral algebraic number, i.e., there are, for some natural n, integers $a_1, \ldots, a_n$ such that $\alpha^n + a_1\alpha^{n-1} + \cdots + a_n = 0$, Dirichlet studied the numbers

$$b_0 + b_1\alpha + \cdots + b_{n-1}\alpha^{n-1}$$

with arbitrary integral coefficients $b_0, \ldots, b_{n-1}$. Let $\mathbb{Z}[\alpha]$ be the ring of all such numbers. Dirichlet saw that, in general, there is no unique decomposition of the numbers in $\mathbb{Z}[\alpha]$ as products of prime elements. (Uniqueness means uniqueness up to multiplication by a unit of $\mathbb{Z}[\alpha]$, i.e., a number $\varepsilon \in \mathbb{Z}[\alpha]$ such that ε^{-1} is also in $\mathbb{Z}[\alpha]$). He proved one of the deepest theorems in algebraic number theory, his theorem about the structure of the group E of all units in $\mathbb{Z}[\alpha]$:

Theorem 3. *E is the direct product of the finite cyclic group of roots of unity in E and a free abelian group of rank $r - 1$, where r denotes the sum of the number of real conjugates and half the number of complex conjugates of α.*

Later on Dedekind understood that the fundamental notion in this connection is the notion of the field $\mathbb{Q}(\alpha)$ generated by α over the rationals, and that in order to generalize the unique factorization into primes one has to consider the ring of all integers of this field. In some cases this ring has the form $\mathbb{Z}[\alpha]$. This is the case for quadratic number fields and for cyclotomic fields $\mathbb{Q}(\zeta)$, where ζ is a root of unity. The rings $\mathbb{Z}[\zeta]$ with ζ of prime order p were studied by Kummer and applied to prove Fermat's last theorem in special cases.

Dirichlet's proof for his unit theorem, which is valid for the ring of integers of an algebraic number field as well, is based on the simple observation that the map which associates to every unit ε the vector in $\mathbb{R}^r$ whose components are the logarithms of the absolute values of the conjugates of ε, defines a homomorphism φ of E into a lattice L in the subspace S of $\mathbb{R}^r$ consisting of the vectors whose components add up to zero. The kernel of φ consists of the roots of unity in $\mathbb{Z}[\alpha]$. This observation shows that the group E has rank $\leq r-1$. The difficult part of the proof consists in the search for $r-1$ independent units, i.e., in showing that the image of φ is a lattice in S. The idea for the solution of this problem came to Dirichlet while hearing the Easter mass in the Sistine Chapel in Rome in 1844.

Summarizing, one can say that important parts of mathematics were essentially influenced by Dirichlet. His proofs characteristically started with surprisingly simple observations, followed by extremely sharp analysis of the remaining problem. With Dirichlet began the golden age of mathematics in Berlin.

References

[1] *Lejeune Dirichlet, P.G.*, Werke, Bd. 1, herausgegeben von L. Kronecker, Bd. 2, fortgesetzt von L. Fuchs, Georg Reimer, Berlin, 1889.

[2] *Biermann, K.-R.*, Johann Peter Gustav Lejeune Dirichlet, Dokumente für sein Leben und Wirken, Abh. der Deutschen Akademie der Wissenschaften zu Berlin, Klasse für Mathematik, Physik und Technik Nr. 2, 1959.

[3] *Biermann, K.-R.*, Briefwechsel zwischen Alexander von Humboldt und Peter Gustav Lejeune Dirichlet, Berlin, 1982.

[4] *Butzer, P.L., Jansen, M., Zilles, H.*, Johann Peter Gustav Lejeune Dirichlet (1805–1859), Genealogie und Werdegang, Dürener Geschichtsblätter Nr. 71, Düren, 1982.

[5] *Minkowski, H.*, Peter Gustav Lejeune Dirichlet und seine Bedeutung für die heutige Mathematik, Jber. dt. Math.-Verein. **14**, 1905.

[6] *Schubring, G.*, Die Promotion von P.G. Lejeune Dirichlet, Biographische Mitteilungen zum Werdegang Dirichlets, NTM-Schriftenreihe für Geschichte der Naturwissenschaften, Technik und Medizin **21** (1) (1984), 45–65.

Carl Gustav Jacob Jacobi

Herbert Pieper

Jaques Simon Jacobi was born on December 10th, 1804, in Potsdam. There he grew up together with his sister, Therese, and with his two brothers, Moses and Eduard, in a Jewish family; his brother Moses became the physicist and electrical engineer known as Moritz Hermann Jacobi. From November, 1816, until April, 1821, he attended the *Gymnasium* in Potsdam, the same school that Hermann Helmholtz attended some years later. Jacobi began his studies at the University of Berlin in 1821. For some time he attended philosophical, philological and mathematical courses, but at the age of nineteen Jacobi decided to concentrate his efforts exclusively on mathematical studies; in doing so, he mainly acquired his mathematical knowledge through private lessons. Jacobi then soon focused on a university career; but he had to convert to Christianity because the edict of 1812, which had given Jews living in Prussia the right to be elected for academic positions, was overruled in 1822 (see [13]). After converting to Protestantism, he changed his first names to *Carl Gustav Jacob.*

In the fall of 1825 Jacobi obtained his doctoral degree with the thesis *Disquisitiones analyticae de fractionibus simplicibus*, which deals with the resolution of algebraic fractions. At the same time he was appointed *Dozent* (lecturer) at the University of Berlin. In his first course he lectured on the analytic theory of curves and surfaces in 3-space. Jacobi's brillant teaching impressed the *Kultusminister* (Minister of Culture) so much that he was given the position of a lecturer (in place of Wrede, who died in 1825) at the University of Königsberg (today Kaliningrad). On December 28th, 1827, he was appointed *Extraordinarius* (associate professor) and on July 7th, 1832, he was promoted to *Ordinarius* (full professor) in Königsberg. Jacobi lived and worked in Königsberg for seventeen years. He interrupted his research and teaching only for official visits to Paris in 1829, to London and Manchester in 1842, for trips to Berlin and Potsdam and for taking the baths in Marienbad in 1839. On September 11th, 1831, Jacobi married Marie Schwinck, the daughter of a wholesale merchant in Königsberg; they had three daughters and five sons.

In 1841 Jacobi became ill with diabetes mellitus. Dirichlet and A. von Humboldt managed to secure a donation from the King of Prussia, which enabled Jacobi to spend nearly one year in Italy convalescing. After residing in Rome and Naples together with his pupil Borchardt and Dirichlet, Jacobi was able to return to Berlin in June 1844. By an Order in Council he was given the permission to stay in Berlin until he recovered completely from his illness. From September 1844 on he lived in Berlin as a member of the Prussian Academy of Sciences. (Jacobi had become a corresponding member of the Prussian Academy of Sciences in 1829 and a non-resident member in 1836, see [9].) He was entitled, but not obliged, to lecture at the University of Berlin. In the spring of 1848 he applied for a professor-

Carl Gustav Jacob Jacobi

ship at the Faculty of Philosophy of the University of Berlin, but this was declined (see [9]). In fact, this decision was in reaction to Jacobi's political activities during the revolutionary year of 1848. In May 1849, again because of Jacobi's political attitude, the *Kultusminister* inquired whether Jacobi's health had improved sufficiently enough so that he could return to Königsberg, and a few days later he was deprived of the permisson given by the King of Prussia to have his residence in Berlin. Jacobi had to give up his home in Berlin. His family took up residence in Gotha (where life was considerably less expensive), while Jacobi moved to a hotel in Berlin. Towards the end of 1849 Jacobi was offered professorships first in Prague and later in Vienna. On January 19th, 1850, Jacobi accepted the offer of the University of Vienna. Meanwhile, A. von Humboldt convinced the King of Prussia to restore Jacobi's residence permit so that Jacobi could remain in Berlin. On February 18th, 1851, Jacobi died; he was buried at the Trinity churchyard next to *Hallesches Tor* in Berlin.

In his research, Jacobi covered a wide range of mathematical topics and was therefore considered to be one of the most creative and most productive mathematicians. At the same time he was a very gifted teacher and initiated reforms in mathematical university teaching. Together with Neumann and Sohncke he founded in 1834 one of the first mathematical-physical seminars. The University of Königsberg became a centre of mathematical research and gained a great influence on other German universities. Although being strongly involved with mathematical research, Jacobi also found the time to deal with mathematical historical questions (see [10], [11]). In addition, he was a brilliant speaker, a highly respected philologist and a notable musician.

Jacobi's wide-reaching research activities are reflected in about 170 publications and two monographs. Almost every branch of mathematics received important contributions from him. F. Klein said that Jacobi's contributions to the theory of elliptic functions are his "most original results" (see [6], p. 109: "*Jacobis originellste Leistungen*"). With the publication of the *Fundamenta nova theoriae functionum ellipticarum* (New foundations of the theory of elliptic functions) in the year 1829, Jacobi was indisputably recognized as being the second ranking mathematician after C.F. Gauss among the German mathematicians. Jacobi made various applications of his newly founded theory of elliptic functions to number theory, to the theory of integration, to geometry, to mechanics and astronomy. His idea of considering the inverse functions of the elliptic integrals could also be applied to hyperelliptic (Abelian) integrals.

More precisely, Jacobi's publications concerned such areas as various parts of number theory (e.g., Jacobi sums, theory of residues, cubic and biquadratic reciprocity laws); analytical mechanics (dynamics) and differential equations (e.g., principle of the last multiplier); and, moreover, algebraic geometry (e.g., plane curves of higher degree, surfaces of the second degree, Poncelet polygons); algebra (e.g., determinants); analysis (e.g., inversion of power series, spherical harmonics and related functions, functional determinants); calculus of variations, differential geometry (e.g., theory of geodesics); celestial mechanics, astronomy (e.g., attractions of ellipsoids, theory of configurations of rotating liquid masses); and the history of mathematics.

Since it is impossible in this short note to describe in detail all the results due to Jacobi, who was sometimes also called the "Euler of the 19th century" (see [1], p. 102), together with their impact on the development of mathematics, we will concentrate on one aspect of Jacobi's scientific legacy.

The triple product identity: Jacobi and the French mathematician A.-M. Legendre corresponded for five years, from August 1827 to June 1832. This correspondence provides insight into the famous mathematical competition between Jacobi and Abel related to the development of the theory of elliptic functions, the inverse functions of the elliptic integrals. The theory of elliptic integrals was created in the 18th century by L. Euler, who initiated this theory after having received a copy of Fagnano's work on December 23rd, 1751, a date which Jacobi later called the birthday of the theory of elliptic functions (see [2], p. 23). Elliptic integrals had also been one of Legendre's favourite objects of study since 1786. He systematized the classical theory of elliptic integrals of various kinds, their addition formulas and transformation theorems during the years 1825/26. Shortly after this period the newly founded theory of elliptic functions appeared, which made much of Legendre's work in this respect obsolete. It has to be noted that there is no indication that either Euler or Legendre thought of considering the inverse functions of elliptic integrals: Euler preferred geometrical, Legendre numerical applications of their theory of elliptic integrals. Moreover, we point out that Legendre seemed to have had an aversion towards the complex numbers, which were crucial for introducing the elliptic functions.

The insight of considering the inverse functions of elliptic integrals is due to Gauss, Abel and Jacobi. Gauss had this idea in the late 1790s, but he did not publish his results. Abel had the same idea in 1823 and published his discoveries in September 1827. Jacobi had the idea during the summer of the year 1827 and published his results in December 1827. But Jacobi's first announcement of his results on this subject already appeared in September 1827. The corresponding two notes are dated June 13th and August 2nd of that year and were printed in the *Astronomische Nachrichten.* They presented special transformation laws and a general formula for elliptic integrals, but neither of the two notes contained any proof. After the September issue of the *Astronomische Nachrichten* had arrived in Paris, Legendre immediately became aware of Jacobi's formulas and, in early November 1827, he received a first letter from Jacobi explaining his new results. Soon afterwards Legendre gave an enthusiastic account of Jacobi's discoveries in a meeting of the Academy of Sciences in Paris. In his publication of December 1827, Jacobi gave complete proofs of his results; he there also introduced the inverse functions of elliptic integrals and established their double periodicity. In September 1827 Abel's first memoir on elliptic functions, his *Recherches sur les fonctions elliptiques*, was published in *Crelle's Journal*. Here we also find the fundamental principle of considering the inverse functions instead of the elliptic integrals themselves and the principle of double periodicity. By means of these results Abel could have derived Jacobi's results mentioned above, and, on the other hand, Jacobi could have read Abel's paper in October 1827! Finally, in April or May of the year 1828, Abel became aware of Jacobi's proof. He was shocked to see Jacobi using the same methods as he had done. Now the famous contest between Abel and Jacobi began. This competition ended abruptly in April 1829 because of Abel's death on April 6th, just two days before the arrival of a letter from Crelle containing his appointment as a professor to the University of Berlin. As a side remark we emphasize here that Abel and Jacobi never met (see [12]), although the contrary has been stated rather often in the literature.

The question whether Jacobi was led to consider the inverse functions of elliptic integrals only after having read Abel's paper of October 1827, has been raised several times in the history of mathematics, and the arguments trying to prove Jacobi's independence have been presented in a partly poor and partly incorrect manner (e.g., consider the Abel biography by Bjerknes or the polemical publications by E. Dühring). Only after the publication of the correspondence between Legendre and Jacobi was it possible to definitely prove Jacobi's autonomy in the discovery of the theory of elliptic functions; for more details we refer to the publications of Bertrand, Gundelfinger (relying on Weierstrass), Krazer, Koenigsberger and Ore (see [14]). In fact, at first, only Jacobi's letters to Legendre were published by Bertrand in 1869; the publication of the remaining parts of the correspondence was completed by Borchardt in 1875. A new edition of the whole correspondence is in preparation and will appear soon in the series of *Teubner-Archiv.* This remarkable correspondence is a very rich source for the history of mathematics around 1830; it contains many mathematical problems together with their solutions and therefore allows a rare insight into the development of mathematical theories, in particular, the theory of elliptic functions.

Jacobi considered elliptic functions not only for their own sake, but he was mainly interested in formulas involving these functions and giving rise to interesting implications in number theory. Jacobi's number theoretical results, which occur in fact for the first time in his correspondence with Legendre, mark the beginning of a new era in dealing with number theoretical problems by using the tools of complex analysis. Here we only mention the celebrated triple product identity, which Jacobi applied to various number theoretical questions, e.g., to determine the number of representations of a natural number as a sum of four squares. This identity, which is also called the two variable Jacobi theta function identity, is given by the following equation:

$$\sum_{n=-\infty}^{+\infty} q^{n^2} z^{2n} = C(q) \cdot \prod_{m=0}^{+\infty} (1 + q^{2m+1} z^2)(1 + q^{2m+1} z^{-2}), \tag{1}$$

where the factor $C(q)$ depends only on the variable q, and has the form

$$C(q) = \prod_{m=1}^{+\infty} (1 - q^{2m}).$$

Jacobi published this identity in his *Fundamenta Nova* in April 1829; in the letter of April 12th, 1828, to Legendre he formulated it in a somewhat different form without giving the factor $C(q)$. The importance of the triple product identity relies on the fact that it relates the representation of the theta function as an infinite series with its representation as an infinite product. Jacobi often pointed out that it was a very hard problem for him to determine the factor $C(q)$, which is independent of the second variable z. As the author showed (see [14]), this factor can in fact be derived from Euler's identity on pentagonal numbers. In a letter of 1848, Jacobi claimed his identity to be perhaps the most important and the most fruitful formula which he found in pure mathematics ("[...] *Formel, welche wohl das wichtigste und fruchtbarste ist, was ich in reiner Mathematik erfunden habe*," see [2], p. 60). Indeed, this identity led Jacobi to results which belong to the most interesting discoveries in mathematics of that period.

We would like to end this discussion by considering the application of the triple product identity, which leads to the above-mentioned result about the determination of the number of representations of a natural number as a sum of four squares: In Jacobi's letter of April 12th, 1828, to Legendre we find the formula

$$\sqrt{\frac{2K}{\pi}} = 1 + 2q + 2q^4 + 2q^9 + 2q^{16} + \dots \ . \tag{2}$$

The proof of this formula is given in Jacobi's *Fundamenta Nova* using the triple product identity. In his subsequent letter of September 9th, 1828, to Legendre, he called this formula one of the most brilliant results of the whole theory and added that everything concerning the representations of numbers as sums of squares is related to the theory of elliptic functions by means of it. On the other hand, in the same letter to Legendre, Jacobi states the following second formula involving K (the elliptic integral of the first kind with modulus k and amplitude $\pi/2$)

$$\begin{aligned} \left(\frac{2K}{\pi}\right)^2 &= 1+\frac{8q}{1-q}+\frac{16q^2}{1+q^2}+\dots \\ &= 1+8\sum_u \sigma_1(u)[q^u+3q^{2u}+3q^{4u}+3q^{8u}+3q^{16u}+\dots]; \end{aligned} \tag{3}$$

here the sum is taken over all odd natural numbers u and $\sigma_1(u)$ denotes the sum of all divisors of u. Comparing formulas (2) and (3) gives the equation

$$(1+2q+2q^4+2q^9+\dots)^4 = 1+8\sum_u \sigma_1(u)[q^u+3q^{2u}+3q^{4u}+3q^{8u}+3q^{16u}+\dots], \tag{4}$$

which in turn leads to the desired formula for the number of representations of a natural number as a sum of at most four squares.

We close this note by citing the following historical anecdote: In 1828 Legendre published a first supplement of his *Traité des fonctions elliptiques* presenting Jacobi's and Abel's recent results on the theory of elliptic functions. Fourier informed the Paris Academy of Sciences about it in his annual report. After apologizing for not being able to go into details, he added the somewhat critical remark that without any doubt these results belong to the most beautiful and deepest ones in mathematics, but that such highly qualified persons should also direct their research interests to applications of mathematical knowledge to solve problems arising from the natural sciences, which is the ultimate measure of progress of human intelligence. In Spring 1829 Jacobi send a copy of his book *Fundamenta Nova* to the Academy of Sciences in Paris and in December of the same year Poisson reported on it quoting the above-mentioned, critical remark of Fourier. After having read Poisson's report, Jacobi wrote in his letter of July 2nd, 1830, to Legendre (see [3], pp. 272, 273): "I read the report of Mr. Poisson on my work with great pleasure. I think, I can be quite happy about it... But in this report Mr. Poisson cited a not very suitable remark of the deceased Mr. Fourier, in which the latter expressed his concern that we, Mr. Abel and myself, did not give the preference to a treatment of the heat flow." And now Jacobi added the following, well-known statement: "It is true that in Mr. Fourier's opinion the principal aim of mathematics is its public utility and the explanation of natural phenomena; but a philosopher such as himself should have known that the only aim of science is the honour of the human mind, and that under this title a question about numbers is worth as much as a question about the system of the world" (see [7], [15]).

References

[1] *Ahrens, W. (ed.)*, Briefwechsel zwischen C.G.J. Jacobi und M.H. Jacobi, Leipzig, 1907.

[2] *Ahrens, W., Stäckel, P. (eds.)*, Der Briefwechsel zwischen C.G.J. Jacobi und P.H. von Fuß über die Herausgabe der Werke Leonhard Eulers, Leipzig, 1908.

[3] *Borchardt, W. (ed.)*, Correspondance mathématique entre Legendre et Jacobi, J. Reine Angew. Math. **80** (1875), 205–279.

[4] *Borchardt, W.* (vol. 1), *Weierstraß, K.* (vols. 2–7), *Lottner, E.* (suppl.) *(eds.)*, C.G.J. Jacobi's Gesammelte Werke, Berlin, 1881/91.

[5] *Jacobi, C.G.J.*, Fundamenta nova theoriae functionum ellipticarum, Königsberg, 1829.

[6] *Klein, F.*, Vorlesungen über die Entwicklung der Mathematik im 19. Jahrhundert, Teil I, Springer-Verlag, Berlin, 1926.

[7] *Knobloch, E., Pieper, H., Pulte, H.*, "... das Wesen der reinen Mathematik verherrlichen", Reine Mathematik und mathematische Naturphilosophie bei C.G.J. Jacobi, Mit seiner Rede zum Eintritt in die philosophische Fakultät der Universität Königsberg aus dem Jahre 1832, Math. Semesterberichte **42** (1995), 1–32.

[8] *Koenigsberger, L.*, Carl Gustav Jacob Jacobi, Festschrift zur Feier der hundertsten Wiederkehr seines Geburtstages, Leipzig, 1904.

[9] *Pieper, H.*, Jacobi in Berlin, in: Die Entwicklung Berlins als Wissenschaftszentrum (1870–1930), Beiträge einer Kolloquienreihe Teil IV, Institut für Theorie, Geschichte und Organisation der Wissenschaften der AdW der DDR, Heft **30** (1982), 1–35.

[10] *Pieper, H.*, Jacobis Bemühungen um die Herausgabe Eulerscher Schriften und Briefe, Wiss. Zeitschrift der Pädagogischen Hochschule Erfurt-Mühlhausen, Math.-Nat. Reihe **20** (1984), 78–98.

[11] *Pieper, H. (ed.)*, Briefwechsel zwischen Alexander von Humboldt und C.G. Jacob Jacobi, Beiträge zur Alexander-von-Humboldt-Forschung, Berlin, 1987.

[12] *Pieper, H.*, Urteile C.G.J. Jacobis über den Mathematiker E.E. Kummer, NTM-Schriftenreihe für Geschichte der Naturwissenschaften, Technik und Medizin **25** (1988), 23–36.

[13] *Pieper, H.*, C.G. Jacob Jacobi (1804–1851), in: Rauschning, D., Nerée, D. v. (eds.), Die Albertus-Universität zu Königsberg und ihre Professoren, Aus Anlaß der Gründung der Albertus-Universität vor 450 Jahren, Berlin, 1995, 473–488.

[14] *Pieper, H. (ed.)*, Korrespondenz zwischen Legendre und Jacobi, Correspondance mathématique entre Legendre et Jacobi, Teubner-Archiv, Leipzig, 1998.

[15] *Pulte, H. (ed.)*, Carl Gustav J. Jacobi, Vorlesungen über analytische Mechanik (Berlin 1847/48), Nach einer Mitschrift von W. Scheibner, Dokumente zur Geschichte der Mathematik 8, Braunschweig/Wiesbaden, 1996.

Jacob Steiner and Synthetic Geometry

Heinrich Begehr and Hanfried Lenz

Jacob Steiner (18.3.1796–1.4.1863) was born in Utzendorf near Bern in Switzerland on March 18, 1796, as son of a farmer. Growing up on a farm, he did not learn to write until his 14th year (see appendix in [5]). He then entered Pestalozzi's pedagogical institute at Iferten (Yverdon), at first as a student, then as a teacher. Here he learned to combine numbers with spatial illustrations. In 1818 Steiner moved to Heidelberg where he lived as a private teacher. There he acquired knowledge of contemporary French geometry. In Heidelberg he received his PhD.

As the Prussian Ministry of Education was interested in Pestalozzi's ideas, Steiner was offered to go to Berlin as a teacher in 1821. Until 1822 he taught at a private school, *Plamannsches Privat-Institut*, and at the *Werdersches Gymnasium*. Afterwards he again lived on private lessons. One of his pupils was the eldest son of Wilhelm von Humboldt (see [B], [K]) and through this connection he met Alexander von Humboldt who had returned to Berlin in 1827 after 20 years in Paris. Alexander von Humboldt and August Leopold Crelle became the main supporters of young scientists in Berlin for more than two decades. Some of the first supported mathematicians were Jacob Steiner, Niels Henrik Abel, Carl Gustav Jacob Jacobi and Gustav Peter Lejeune Dirichlet. In the mid 1820s one could see Crelle, Abel and Steiner walking through the city and people called them "Adam and his two sons, Cain and Abel" (see [B]).

Steiner and Dirichlet were the first important mathematicians after Euler and Lagrange to be elected as members of the Prussian Academy of Sciences. Both faced problems in becoming full professors at the university and, in fact, Steiner never made it. He even had problems obtaining the licence to teach at public schools. In the qualifying examination he was very weak in philosophy, languages and history; only in mathematics he was considered adequate. He showed a deep knowledge of elementary geometry but he did not do well in analysis, in algebra and in spherical geometry (see [B], p. 56). Therefore he was granted the licence to teach only middle school classes. Only in 1826, when he repeated the examination, did he succeed and in 1829 he became a regular teacher (*Oberlehrer*) at a high school. From 1825 till 1835 Steiner also taught students of the *Gewerbeschule*, one of the predecessors of the Technical University of Berlin. Here he had problems because of his rude behavior toward students. In winter time he sometimes opened the windows of the classroom hoping the students would quit (see [B], pp. 56, 57). Or he talked about his time as a herdsman, when for days he would be alone in the mountains with only his herd. And fixing his gaze on his students he assured them that he had learned to recognize any of his oxen, even at long distances. But at the university where he became *Extraordinarius* (associate professor) on October 10, 1834, his influence on students was tremendous. Many students being trained as teachers preferred his kind of teaching because he was asking questions

Jacob Steiner

and asking them to perform constructions. Sometimes he closed the curtains in order to darken the classroom so that the students could better follow and "see" his geometric constructions which he was describing using words and sometimes his hands.

It was only when he was 30 years old that he started research. He became famous as a founder of synthetic geometry. In 1832 the University of Königsberg honoured him with a PhD degree and in 1834 he became a member of the Prussian Academy of Sciences. The reasoning in his papers was purely geometric. He was able to see geometrical connections which no one else could. And he disliked analytical argumentation, considering it superfluous because of his intuitive understanding of the geometrical situation. He once said: "I can see that this is true and Jacobi says he can even prove it." Many of his theorems were published without any proof and this made it difficult for Weierstrass later to edit the collected works of Jakob Steiner. One of his pupils was Heinrich Schröter who became the successor to Kummer in Breslau.

Already as a private teacher Steiner began his investigations on circles and spheres [2], which were published more than 100 years later, in 1931. He was a regular author of Crelle's famous *Journal für die reine und angewandte Mathematik*. In its first volume (1826) he published his paper [6] *Einige geometrische Betrachtungen* which deals with circles (power line of two circles, points of similarity, problems of Apollonius and Malfatti, and many other topics). After his death on April 1st of 1863, this paper was published as a booklet in the series of *Ostwalds Klassiker*. Steiner's most famous work is [3] *Systematische Entwicklung der Abhängigkeit geometrischer Gestalten voneinander...* He planned to publish a big monograph of five parts under this title, but only the first appeared, in 1832, and again much later as two booklets in *Ostwalds Klassiker*. The basic idea is the following projective generation of a conic: Let P be a pencil of lines in the projective plane, and Q its image under a projective transformation f; then the intersections of corresponding lines form a conic or a line, and each conic can be generated thus. This construction of conics immediately implies Pascal's theorem, and Brianchon's dual theorem (see [4], vol. II, p. 97). It allows one to draw arbitrarily many points and tangents of the conic determined by five given points (or by five given tangents), using only a ruler. Steiner also extends his construction to 3-space, where ruled quadrics are generated analogously.

As a consequence of Pascal's theorem, Steiner studied the 60 Pascal lines of six points on a conic. They are partitioned into 20 triples each of which has a point in common which is called a Steiner point (see [4], vol. II, pp. 128–137).

Some main ideas of synthetic geometry were already present in Poncelet's work of 1822, which consolidated a collection of special theorems into a distinct branch of geometry, namely, projective geometry. But Steiner's work established the synthetic method in projective geometry as well as his personal fame as a mathematician, especially in Germany (see [K]).

Synthetic geometry, and in particular Steiner himself (see [4]) tried to avoid algebra, such as calculations with coordinates, as much as possible. But analytic geometry, that is, the use of algebra in geometry, certainly is much more powerful. On the other hand, synthetic or "pure" reasoning was often more elegant than algebraic methods, especially before the rise of abstract algebra. Synthetic geometry works well in the study of quadratic curves and surfaces, but already for degree three the use of algebra is indispensible. For the history of the opposing synthetic and analytic schools, we refer to [K].

In 1833 Steiner published his little book [5] on geometric constructions with the ruler and one fixed drawn circle, later reprinted in *Ostwalds Klassiker* Nr. 60. This latter edition also contains an appendix with a more detailed biography of Steiner. The basic idea of this paper had already appeared in Poncelet's book of 1822.

Steiner's collected mathematical papers (*Werke*) contain a huge amount of geometric problems and theorems. Very often he gave no proofs of his theorems nor did he quote earlier sources by other mathematicians. As an example, look at the paper of 1848, *Allgemeine Eigenschaften der algebraischen Kurven*, [1], vol. II, pp. 493–500, where Steiner failed to mention that most of his results were due to earlier work by Plücker (see [K], p. 128).

In the following, some geometric theorems of Steiner are presented which can be stated in a few words. A complete quadrilateral contains four triangles. Their circumscribed circles have a point in common which is on the circle through their centers (Steiner circle), and their orthocenters (intersection points of the three altitudes) lie on a line ([1], vol. I, pp. 128, 233).

Let $ABCD$ be a quadrangle inscribed in a circle K. Then the orthocenters of the triangles ABC, ABD, ACD, and BCD lie on a circle congruent to K ([1], vol. I, p. 128).

Let S be the intersection point of AB and CD. Then the bisectors of the angles ASD and ATD are parallel ([1], vol. I, p. 128).

n circles divide the plane into at most $nn - n + 2$ parts with one of them infinite ([1], vol. I, p. 83). Steiner gave analogous theorems on n lines dividing the plane and n planes dividing space ([1], vol. I, pp. 81 and 87, 88).

If two quadratic surfaces are tangent in more than two points then they are tangent in the points of a plane conic ([1], vol. I, p. 9).

Three circles in the plane have six similarity points. There are four lines containing three of these points each ([1], vol. I, p. 29). Let K and L be two disjoint circles in the plane, and let $A, B, C, \ldots$ be circles tangent to K and L such that, moreover, B is tangent to A, C tangent to B and so on. If the n-th circle of this sequence coincides with A then this property is independent of the choice of A ([1], vol. I, pp. 43, 135).

Let three lines in the plane be tangents of two conics. Then the six tangent points are on a third conic ([1], vol. I, p. 134). Let four lines be tangents of two conics. Then the eight tangent points are on a third conic ([1], vol. II, p. 427).

In a convex polygon there are in general half as many inside intersection points of diagonals as outside intersection points of non-adjacent sides ([1], vol. I, p. 177).

Mathematics is indebted to Steiner for many special results and problems in different fields.

A well-known theorem of Steiner relates the moments of inertia for two parallel rotation axes, one of them through the center of gravity.

Also famous is Steiner's symmetrization of a convex body which he used for his proof of the isoperimetric property of the circle, and also the analogous property of the sphere, provided the minimum exists, which, however, Steiner could not yet prove (see [6], [8]; one of the modern existence proofs uses Blaschke's selection theorem of 1916). This symmetrization has become a standard tool of convex geometry.

There are some concepts, problems, and theorems which are connected with Steiner's name although his real share in their origin is small.

Let us begin with Steiner systems, in particular the Steiner triple systems, a combinatorial concept about which there is a huge literature. In a short note *Combinatorische Aufgabe* of 1853 in *Crelle's Journal* (see [1], vol. II, pp. 436–438), Steiner posed a problem. In modern language it reads as follows. Find finite sets of v points and b blocks (i.e., point sets) such that – for given natural numbers t and k (larger than t and smaller than v) – any t points are in exactly one block and each block consists of exactly k points.

If $t = 2$ the blocks are also called *lines*. In Steiner's triple systems one has $t = 2$ and $k = 3$. The two simplest examples are the projective plane of order 2 with 7 points and 7 lines, and the affine plane of order 3 with 9 points and 12 lines. The latter is also well-known as the configuration of the 9 complex inflection points of a nonsingular plane algebraic curve of degree 3 (only three of them can be real), which were discovered by Plücker in 1839. It is easy to see that a Steiner triple system can only exist if v is congruent modulo 6 to 0 or 1.

When Steiner formulated his problem, W.S.B. Woolehouse had already introduced the triple systems (1844), and T.P. Kirkman had already proved that the just-mentioned necessary condition is indeed sufficient for the existence of a Steiner triple system with v points. Thus one may doubt whether naming the systems after Steiner was appropriate; at any rate it is now firmly established. The most famous Steiner system is the so-called large Witt design, where $t = 5$, $k = 8$, $v = 24$, $b = 759$. Its automorphism group (with the obvious definition) is the 5-fold transitive large Mathieu group of order

$$48 \times 24 \times 23 \times 22 \times 21 \times 20 = 244823040 \ ,$$

one of the earliest sporadic simple groups. Nowadays one knows a great deal about Steiner systems for $t = 2$, something for $t = 3$, a little for $t = 4$ and $t = 5$, and nothing at all for $t = 6$ or more. No Steiner system is known in this case, nor is there a known non-existence proof. In modern combinatorial theory the Steiner systems have been generalized to the so-called t-designs, about which there is a large and fast growing body of literature.

Also the expressions Steiner problem, Steiner network, and Steiner tree are in common use. Three or more cities (ideally vertices of a triangle) are to be connected by a network of streets (ideally segments of lines) of minimal length. If all the angles of the triangle are smaller than 120 degrees, then the solution is given by three segments which meet in a point inside the triangle at angles of 120 degrees. Theoretically the problem for n points is simple, at least in the plane. In general there appear additional points, so-called Steiner points where three segments of the network meet under 120 degrees. But the complexity of the practical solutions is high, at the present state of knowledge the amount of time needed grows exponentially with n (technically speaking, the problem is NP-hard). Again the problem, on which there is quite a lot of recent literature, has many fathers besides Steiner, some of them much older such as Torricelli and Cavalieri.

The Steiner (Roman) surface is a special surface of fourth degree with a threefold point which intersects each of its tangential planes in two conics. For this surface – which Steiner investigated without publishing anything on it – we refer to the posthumous paper in [1], vol. II, pp. 723, 724, and the commentary of Weierstrass in [1], vol. II, p. 741.

In this final remark (see [1], vol. II, pp. 742, 743), Weierstrass points out that Steiner was human and hence not free from errors. For details see the comments in [1], vol. I, pp. 523–527, and [1], vol. II, pp. 727–743.

References

[1] *Steiner, J.*, Gesammelte Werke I, II, ed. Karl Weierstrass, Georg Reimer, Berlin, 1881, 1882.

[2] *Steiner, J.*, Allgemeine Theorie über das Berühren und Schneiden der Kreise und Kugeln, Posthumous Edition by K. Fueter and F. Gonseth, Orell Füssli Verlag, Zürich und Leipzig, 1931.

[3] *Steiner, J.*, Systematische Entwicklung der Abhängigkeit geometrischer Gestalten voneinander mit Berücksichtigung der Arbeiten alter und neuer Geometer über Porismen, Projektionsmethoden, Geometrie der Lage, Transversalen, Dualität und Reziprozität, etc., Fircke, Berlin, 1832 (see also [1], vol. I, 229–460, or Ostwalds Klassiker der exakten Naturwissenschaften 82, 83, Engelmann, Leipzig, 1896).

[4] *Steiner, J.*, Vorlesungen über synthetische Geometrie I, II, Teubner, Leipzig und Berlin, 1867.

[5] *Steiner, J.*, Die geometrischen Konstruktionen, ausgeführt mittels der geraden Linie und eines festen Kreises, Dammler, Berlin, 1833 (see also [1], vol. I, 461–522, or Ostwalds Klassiker der exakten Naturwissenschaften 60, Engelmann, Leipzig, 1895).

[6] *Steiner, J.*, Einige geometrische Betrachtungen, J. Reine Angew. Math. **1** (1826), 161–184, 252–288 (see also Ostwalds Klassiker der exakten Naturwissenschaften 123, Engelmann, Leipzig, 1901).

[7] *Steiner, J.*, Über Maximum und Minimum bei den Figuren, in der Ebene, auf der Kugelfläche und im Raume überhaupt, J. Reine Angew. Math. **24** (1842), 93–162, 189–250 (see also [1], vol. II, 177–242, 245–308, or J. Math. Pures Appl. Ser. 1, **6** (1841), 105–170).

[8] *Steiner, J.*, Einfache Beweise der isoperimetrischen Hauptsätze, J. Reine Angew. Math. **18** (1838), 281–296 (see also [1], vol. II, 77–91).

[B] *Biermann, K.-R.*, Die Mathematik und ihre Dozenten an der Berliner Universität 1810–1933, Akademie Verlag, Berlin, 1988.

[K] *Klein, F.*, Vorlesungen über die Entwicklung der Mathematik im 19. Jahrhundert, 2 vols., Springer-Verlag, Berlin, 1926–1927.

Gotthold Eisenstein
16 April 1823–11 October 1852

Norbert Schappacher

The early development and untimely death of a genius is a well-known theme in general ("Whom the gods love ..."), and in musical biography in particular, with Mozart and Schubert as outstanding examples. But three geniuses among the nineteenth century mathematicians, whose lives were cut off all too soon, make Mozart's death at close to 37 years, and even Schubert's at hardly less than 32 years, almost seem to have come at a reasonably mature age. The mathematicians are Evariste Galois – who lost his life at 20 in an absurd duel, Niels Henrik Abel – who succumbed to tuberculosis at age 26, and finally Gotthold Eisenstein, whose frail body held out exactly 1000 days longer than Abel's, before giving in to the same disease.

If it is true that Eisenstein died young, this still did not spare him a few miserable years during his life – nor even a dubious afterlife. In fact, few people loved him. The scientific world did bestow fantastic honours on the very young man. But the mathematical community did not keep an adequate memory, either of his person, or of his visionary work. Felix Klein for instance – in his exaggerated endeavor to elevate Riemann beyond comparison – portrayed Eisenstein not only as megalomaniacal and paranoid, but also mathematically as a *Formelmensch*, a man of formulas (see [4]). This was an expression of the Göttingen spirit from the turn of the century, under Hilbert and Klein, a spirit that contrasts sharply with Gauss's unusually favourable attitude towards Eisenstein. (Hilbert, although a little more favourably disposed towards Eisenstein's œuvre than Klein, strove in his grandiose systematic treatment of algebraic number theory, the *Zahlbericht* of 1897, to replace what he called Kummer's calculations in the arithmetic of cyclotomic fields by "Riemann's conceptual method." He was thus proud to show that he could do completely without Kummer's logarithmic derivatives of units, i.e., without a technique that has since come to be a key tool of contemporary arithmetic through the work of Artin, Hasse, Iwasawa, Shafarevich, Coates and Wiles.)

But let us take things in order.

The very fact that Ferdinand Gotthold Max Eisenstein, the first of six children, all born in Berlin, survived childhood can be seen as an exception, for all his siblings were not so lucky. He survived meningitis and other numerous illnesses. He did remain hypochondriac, though, throughout his life.

His parents were Protestants after having converted from Judaism. The father had served for 8 years in the Prussian army, and then tried to be a businessman, apparently with not much success. Eisenstein was very closely attached to his mother who seems to have had a creative pedagogical instinct. For instance, in one

Ferdinand Gotthold Max Eisenstein

of the two autobiograpical sketches of Eisenstein that we have (the one written in August 1843 and reproduced in [3], predating a shorter one from March 1847 [2]; both texts reveal a curious mixture of naive openness and conventional style), he relates that his mother taught him the letters of the alphabet by associating a pictorial meaning to each of them, like a doorway for "O", and a key for "K".

Intellectually precocious, the child showed as much thirst for knowledge as lack of skill and acculturation in practical everyday matters. His enormous mathematical talent was noticed and encouraged by his teachers in secondary school. He started learning calculus from reading Euler and Lagrange on his own when he was about 15 years old. In 1842 he bought his own copy of Gauss's *Disquisitiones Arithmeticæ*, in French translation. Number theory quickly became his favourite interest in mathematics. During his time at the *Friedrich-Werdersches Gymnasium*, from 1840 to 1842, he also attended, as an 18- and then 19-year-old, courses at the University of Berlin, in particular, some given by Dirichlet.

Until 1842, Eisenstein lived either in or not far from Berlin. In the summer of this year, however, before finishing the *Gymnasium*, he went with his mother on the only big voyage of his life: to England, to join his father who was doing business there at the time. The family seems to have toured England, Ireland and Wales quite a bit, the son reading Gauss all the way, and playing the piano when possible. He would have liked to settle in Dublin. He had met Hamilton there, who entrusted him with a mansucript for the Berlin Academy. The family returned to

Berlin in 1843; the parents separated; and the 20 year old Eisenstein obtained his graduation from school, enrolling as a student at the University of Berlin. The following year he stepped into the limelight of science.

When Eisenstein delivered Hamilton's manuscript to the Academy, he also added a paper of his own, on cubic forms. This led to Crelle's interest in the prodigy. What happened next is not only a proof of Eisenstein's unusual talent, but also an amazing illustration of what the Berlin scientific establishment of the time was capable of doing for a young genius.

Volume 27 of *Crelle's Journal* (1844) contains no less than fifteen papers (and one problem) by Gotthold Eisenstein; the following volume 28 (still 1844) another eight (plus another problem), in German, French, and also Latin. Thus, within the first year, the young author had more publications to his credit than years of age.

In March 1844, Eisenstein was invited to meet Alexander von Humboldt. Humboldt was to remain Eisenstein's lifelong devoted benefactor. Through Humboldt, and partly through Crelle, Eisenstein received, as of April 1844, financial aid from the King, and sometimes from the Academy. This allowed him to survive, but since the payments were never turned into a regular salary, it also left him with an increasing feeling of material insecurity, especially after the 1848 upheaval. Humboldt's various attempts over the years to secure a professorship at some university for the young man did not work out.

In June 1844, Gauss expressed high praise for Eisenstein's papers. That same month, Eisenstein went to Göttingen for two weeks to see Gauss, with a letter of recommendation from Humboldt. No other young talent was ever treated as well as Eisenstein by the Göttingen master. When proposing Dirichlet for the order *pour le mérite* in 1845, Gauss noted he would have almost proposed Eisenstein instead. In 1847, Gauss wrote a foreword to a collection of Eisenstein's papers which was published as a book.

In February 1845, Eisenstein received an honorary doctorate from the University of Breslau.

The tide began to turn around the end of 1845 and the beginning of 1846, at first in a psychological sense: Eisenstein was essentially lonely. Kronecker with whom he met almost daily earlier in 1845, left Berlin. Other friends of comparable age who did stay in Berlin seemed to fade away. The friendly correspondence with Max Stern in Göttingen (the first non-converted Jew to hold a professorship in mathematics in Prussia) could not make up for the lack of direct contact.

Then there were professional problems: Jacobi, with whom Eisenstein had an intense relationship with many ups and downs, started a public priority dispute with Eisenstein in early 1846. From Jacobi's point of view, this bang on the head of an overrated young star had a beneficial effect. For Eisenstein it probably felt like the end of youth. Onlookers like Gauss and the astronomer Schumacher discussed in their correspondence possible effects of mathematical genius on human relations. A modern expert like André Weil finds it "incomprehensible that Jacobi, Cauchy, and Eisenstein have each published proofs" of the quadratic reciprocity law which were "virtually identical with" Gauss's sixth proof, "and that they even raised priority questions among each other about this."

Eisenstein's production continued through his death, totalling some 40 mathematical articles and more than 700 printed pages. In 1846, Humboldt succeeded in freeing Eisenstein from military service. In 1847, Eisenstein obtained his *Habilitation*, i.e., the right to teach courses at the University of Berlin. This he did apparently with good success, drawing between 10 and 20 students to his analysis courses, and around 5 to more advanced topics such as elliptic functions.

In 1851 for the first time, his health did not permit him to teach. In this same year, upon Gauss's proposal, Eisenstein was elected (along with Kummer) corresponding member of the Göttingen Academy. The following year, five months before Eisenstein's death, Dirichlet arranged for him to be elected as member of the Berlin Academy, succeeding Jacobi.

Eisenstein's attitude to the republican ideas that erupted in March 1848 is not transparent. In 1850, Schumacher wrote to Gauss that Eisenstein was "very red", quoting alledged violent anti-reactionary statements of Eisenstein's from as late as the end of 1849. Eisenstein did drop in at Republican Clubs before March 1848, and during the street fighting on March 19, 1848, he was found in a house (not his own) from which snipers had fired on the royal troups. He was arrested, severely maltreated, and violently forced to march, with lots of other prisoners, the long way to the Spandau Citadel from which he was released the day after. He wrote down a very detailed description of what happened to him for a collection of similar reports by other victims of the transport to Spandau. This collection was published as a book by Adalbert Roerdansz in 1848 ([5]): "A Martyrium for Freedom," under the motto *facta loquuntur* (the facts will speak). The declared goal of the book was to accuse the military of abuse of power in order to create awareness and thus eventually improve the conduct of the troops. The fact that Eisenstein agreed to cooperate with this more idealistic than revolutionary project, does not really reveal his political ideas. In his report, he depicts himself as a peaceful teacher at the university who had not done anything unlawful and who found himself in the house where he was arrested simply because he was seeking shelter from the fighting.

Looking back from today's vantage, Eisenstein's mathematics appear to us more up-to-date than ever. It is not so much the harvest of theorems, nor the creation of full-fledged theories, but the way of looking at things which amazes us – and which may have also been what impressed Gauss.

Three major themes pervade Eisenstein's work: first, the theory of forms, where the goal was to generalize Gauss's arithmetic theory of quadratic forms contained in the *Disquisitiones Arithmeticæ.* Already Hilbert in his *Zahlbericht* made the comment, which is commonplace today, that we would have to translate these results into the language of cubic number fields in order to appreciate them.

Second, the theory of higher reciprocity laws, i.e., the search for a generalization of quadratic reciprocity which Gauss had also treated in his *Disquisitiones Arithmeticæ*, and which he continued to come back to, giving new proofs of it. In order to tackle the general reciprocity law for m-th power residues, one has to work in the ring $\mathbb{Z}[\zeta_m]$ of m-th roots of unity, and thus Eisenstein was among the first mathematicians to use Kummer's arithmetic theory of ideal numbers for these rings, which was created in the early 1840s. In one of his last papers (from 1850), Eisenstein managed to settle the most general reciprocity law in the special case where one of the two numbers to be compared is a rational, the other a cyclotomic integer. This is still quite far from the goal, but was crucial progress at the time when it was obtained. It took Kummer a decade longer to go all the way, and Hilbert in his *Zahlbericht* still uses Eisenstein's law as a crucial stepping stone for both proofs of the general reciprocity law that he presents.

Incidentally, when working on this *Zahlbericht*, Hilbert and Minkowski deplored that the collected papers of Eisenstein's had not been edited. But they did not take action themselves, and so it fell finally to the Chelsea Publishing Company to publish the two volumes in 1975. There was a second edition already in 1989.

The third big mathematical theme taken up by Eisenstein was the rapidly developing theory of elliptic functions, which Gauss had studied without publishing work on the subject. Following Fagnano, Euler and Legendre, Abel had made major progress, and Jacobi had set up a whole theory of the domain in his *Fundamenta Nova*. In a major paper of 1847: *Genaue Untersuchung der unendlichen Doppelprodukte* ..., Eisenstein developed his own independent analytic theory of elliptic functions, based on the technique of summing certain conditionally convergent series. André Weil made this and Kronecker's subsequent work along these lines the subject of his Princeton Lectures in 1974 which later appeared as a book: "Elliptic Functions according to Eisenstein and Kronecker," [6], which has and continues to have major immediate impact on current research in arithmetic algebraic geometry.

There is another circle of ideas in which Eisenstein's investigations appear to have been visionary from our point of view. Gauss, in the introduction to the seventh chapter on cyclotomy of the *Disquisitiones Arithmeticæ*, mentions that one may develop an analogue of cyclotomy for the lemniscate, i.e., an arithmetic study of the division values of the inverse of the elliptic integral $\int \frac{\mathrm{d}x}{\sqrt{1-x^4}}$. This was actually treated by Gauss himself in unpublished notes, and independently by Abel. Eisenstein developed Abel's theory further in the arithmetic direction, and managed to deduce from it the biquadratic reciprocity law. This may not seem such a big achievement in that this kind of analysis of reciprocity laws in terms of elliptic functions does not go very far towards the general reciprocity for m-th power residues. But Eisenstein's development of the arithmetic of the division of the lemniscate, especially in his long and detailed series of articles from 1850, *Über die Irreductibilität* ..., is startling for the way in which he proceeds. He derives a formula for the multiplication by a prime on the elliptic curve $y^2 = 1 - x^4$ which, through its natural generalization in Kronecker's work, is the direct source of what today is known as the Shimura-Taniyama congruence relation in the theory of complex multiplication. By the way, it is here, in proving the irreducibility of the lemniscatic division equation, that Eisenstein establishes his well-known irreducibility criterion.

These articles *Über die Irreductibilität ...* have also recently led A. Adler [1] to suspect that Eisenstein had some sort of knowledge about Jacobian varieties of Fermat curves.

On a different note, Eisenstein sketched on 7 April 1849, writing into his copy of the *Disquisitiones Arithmeticæ*, a proof of the functional equation for the Dirichlet L-function modulo 4 in the critical strip. It is tempting and plausible to think, as Weil proposed [7], that Riemann received basic ideas for his fundamental paper on the zeta function in conversations with Eisenstein that same month in Berlin.

The final scene shows the 83-year-old Alexander von Humboldt following Eisenstein's coffin at the cemetery at *Blücherplatz*. He had obtained money from the King for Eisenstein to go to Sicily for a cure, but it was already too late. The plague of the nineteenth century had taken yet another distinguished victim.

References

[1] *Adler, A.*, Eisenstein and the Jacobian varieties of Fermat curves, Rocky Mountain Journal of Mathematics **27** (1997), 1–60.

[2] *Biermann, K.-R.*, Eine Selbstbiographie von G. Eisenstein, Mitt. Math. Ges. DDR, Heft **3–4** (1976), 150–154.

[3] *Eisenstein, G.*, Mathematische Werke, 2 vols., Chelsea, 1975.

[4] *Klein, F.*, Vorlesungen über die Entwicklung der Mathematik im 19. Jahrhundert, Teil I, Springer-Verlag, Berlin, 1926.

[5] *Roerdansz, A.*, Ein Freiheits-Martyrium, Gefangene Berliner auf dem Transport nach Spandau am Morgen des 19. März 1848, Protocollarische Aussagen und eigene Berichte der 91 Betheiligten, als Beitrag zur Geschichte des Berliner Märzkampfes gesammelt und herausgegeben von Adalbert Roerdansz, Berlin, 1848.

[6] *Weil, A.*, Elliptic functions according to Eisenstein and Kronecker, Ergeb. Math. 88, Springer-Verlag, Berlin, 1976.

[7] *Weil, A.*, On Eisenstein's Copy of the *Disquisitiones*, in: Coates, J., Greenberg, R., Mazur, B., Satake, I. (eds.), Algebraic Number Theory – in honor of Kenkichi Iwasawa, Advanced Studies in Pure Mathematics 17, Academic Press, Princeton, 1989, 463–469.

Kummer and Kronecker

Harold M. Edwards

Ernst Eduard Kummer (1810–1893)

Ernst Eduard Kummer was the senior member of the triumvirate – Kummer, Weierstrass, Kronecker – that dominated mathematics in Berlin, and in Germany, during much of the second half of the 19th Century. He probably is best known today for an incident that never happened.

Kurt Hensel told of the alleged incident in a 1910 lecture observing Kummer's 100th birthday [10]. In Hensel's story, Kummer gave Dirichlet a manuscript containing what he believed was a proof of Fermat's last theorem some time in the early 1840s. Dirichlet studied it and returned it in a few days, saying that the proof would be correct if cyclotomic integers could be factored into primes in just one way, but that he doubted that this was true.

The story is a wonderful parable for young researchers. When Kummer realized that the assumption of unique factorization was wrong, he did not despair but instead tackled the difficult problem of factorization in cyclotomic fields, developed his famous theory of "ideal complex numbers," and proved Fermat's last theorem for a large class of exponents, thereby earning a significant place in the history of mathematics.

The story is wrong (see [5]) in the main particulars – Kummer was working on reciprocity laws not Fermat's last theorem, he submitted his paper to the Berlin Academy not to Dirichlet, his error was not that he naively assumed unique factorization, and the error he did make was pointed out by Jacobi not Dirichlet – but the point of Hensel's parable is undiminished by these corrections. The young Kummer was indeed foolishly optimistic, submitted a paper proving an impossibly simple and general theorem, and had the strength and tenacity to overcome his chagrin and turn the original setback into a major victory.

Kummer, like Weierstrass, was educated as a secondary school teacher, not a university professor. He was born in 1810 in the Silesian town of Sorau, the son of a medical doctor ([10], [14], [1], [2]). His father died in the epidemic of typhus brought by the return of Napoleon's defeated army from Russia in 1813. His widowed mother, through hard work and thrift, provided good educations to both Ernst Eduard and his older brother, sending them first to the *Gymnasium* in Sorau and then to the university in Halle. She hoped he would study theology and become a clergyman, but readily assented to his wish to become a mathematics teacher instead.

He completed his doctoral degree and passed the necessary state examinations in Halle at the age of 21 and, after a *Probejahr* in his native Sorau, taught for 11 years at the *Gymnasium* in Liegnitz. Only his great talent and deep motivation can account for his many achievements and publications in those years. Years

Ernst Eduard Kummer

later he said, "In my first years in Liegnitz, when I was burdened with 24 teaching hours, of which almost none were in mathematics, I achieved at least as much as I did later under what were outwardly the most favorable circumstances. ([10])" If this fact surprised Kummer in his later years, how much more must it surprise us today when researchers are accustomed to so much lighter teaching loads than in Kummer's time.

But Kummer's achievements in Liegnitz were not in research alone. Judging by the testimony of his university students in later years (cf. [16]), he was probably an inspiring and effective *Gymnasium* teacher for all the students, not just the talented ones. But two of his talented students in Liegnitz went on to have significant careers as mathematicians, Ferdinand Joachimsthal (see [1], p. 50) and Leopold Kronecker. Joachimsthal's name would probably be better known today – he was Kummer's successor in Breslau – had he not died at the age of 43. Kronecker's name is of course universally known.

Kummer's research during this period was in the theory of functions, particularly the theory of the hypergeometric series. On the basis of it, he was made a corresponding member of the Berlin Academy in 1839 and then, in 1842, won appointment to a professorship in Breslau. When he learned that the appointment in Breslau was about to happen, he decided on a radical change in the direction of his research. "I am sitting at home and working very energetically in order to have something for a dissertation for my *Habilitation*, and I am beginning something entirely new to me, namely, the cubic residues of primes of the form $6n+1$. ([10])"

This study of reciprocity laws, inspired by the work of Gauss, and, more immediately, of Jacobi, led to Kummer's most famous achievement, the creation of the theory of "ideal prime factors." Kummer's theory is usually described today as an early version of what has been known since the work of Dedekind as the "theory of ideals," but readers who approach it from this point of view will find it difficult to understand. Kummer was a serious student of philosophy, and there are indications that he seriously pondered the philosophy of mathematics, although he did not publish any writings on that subject. As is clear from the brief notice *Zur Theorie der complexen Zahlen* in which he introduced his ideal prime factors, he gave long and serious thought to the question of how to conceive and how to define these new entities. His choices reflect an algorithmic approach to mathematics like the one later championed by Kronecker, and is far removed from the Dedekindian approach so familiar today.

The first fruit of Kummer's new theory was a pair of conditions on a prime p that imply the truth of Fermat's last theorem for the exponent p. Soon thereafter, he proved that in fact his two conditions are equivalent to a single condition involving the Bernoulli numbers mod p, a condition that can be tested by straightforward computation. Up to that time, Fermat's last theorem had been proved only for the prime exponents 3 (Euler), 5 (Legendre and Dirichlet), and 7 (Lamé), the last two steps having been taken with great difficulty, using methods that showed little promise for higher exponents. Kummer's Bernoulli number condition and existing tables of the Bernoulli numbers proved the theorem at one blow for all primes less than 62 except 37 and 59.

In later years, Kummer was able to prove broader sufficient conditions and, in particular, to prove Fermat's last theorem for all prime exponents less than 100, but this was never his main goal. His main goal was to generalize the law of quadratic reciprocity to higher powers. In 1847 he guessed the form the higher reciprocity law should take – the very statement of his law required his new ideal prime factors – and proceeded to verify it by extensive and sophisticated calculations, but more than ten years were to pass before he was able to prove that the law was true. His treatise of 1859 giving the full proof crowned his efforts in number theory with success and, with minor exceptions, brought these efforts to a close. Most of his later research lay in a different direction, marking the third phase of his research.

It should be told that Kummer's theory of ideal prime factors contained a serious gap (cf. [4]) that remained undetected for the first 10 years of its existence. The proof of a crucial lemma regarding the solutions of a certain congruence is incomplete in the original 1847 version, and only with the publication of a paper in 1857, giving what Kummer called a "clarification and a more complete foundation," was the theory he had been using so successfully given a satisfactory basis.

Great as the success of Kummer's work in function theory had been, the success of his new work in number theory, begun when he moved to Breslau, was far greater. In 1855 when Dirichlet left the University of Berlin to become Gauss's successor in Göttingen, the choice of Kummer to succeed him in Berlin was clear. Not only did Kummer's reputation as a researcher put him at the head of his generation, his abilities as a teacher were also highly regarded.

His style ([16]) as a teacher differed from that of many of his contemporaries in that his university courses rarely touched on material related to his research. By all accounts he was a popular and effective teacher – there are many testimonials to his charm and wit as a lecturer – but his goal was to give his students a firm grounding in the basics and a good orientation in mathematics as a whole, rather than introduce them to research. The introduction to research was reserved for the mathematical seminar he founded with Kronecker and Weierstrass.

Kummer was a devoted husband and father. His first marriage, in 1840, was to Ottilie Mendelssohn, a granddaughter of the philosopher Moses Mendelssohn, and a first cousin of the composer Felix Mendelssohn-Bartholdy and his sister Rebecca, who was the wife of Dirichlet. This match, between a provincial *Gymnasium* teacher and a member of one of Berlin's most cultured and sophisticated families, indicates that Kummer's status was rising even before he became a university professor. Ottilie's death from a sudden illness in 1848 was a heavy blow, but Kummer soon remarried, and when he died he was survived by no fewer than nine children ([14]), not to mention grandchildren and great grandchildren.

While Kummer's work in number theory was utterly devoid of geometry, his research in the third phase of his career was dominated by geometry. During this period he studied what is now called the "Kummer surface" in three-dimensional space and continued the work of Jacobi and Hamilton on ray systems and geometric optics. Kummer's name is known today to students of geometric optics who know nothing of his work in number theory.

In addition to teaching and research, Kummer was active in the administration of the university, where he served terms as both Dean and Rector, and of the Academy, where he was perpetual secretary from 1863 to 1878. Lampe in his eulogy says that Kummer regarded administrative activities as a useful distraction for scholars that gave them the opportunity to gather their forces for their scientific work. He retired from teaching in 1884 at the age of 74, saying that he felt his abilities had begun to decline ([2]), although his colleagues had not noticed any impairment. After 9 years of placid and relatively healthy retirement, he died of influenza in 1893.

Leopold Kronecker (1823–1891)

In spite of Kronecker's immense achievements in diverse fields of mathematics – number theory, algebraic geometry, the theory of elliptic functions, Galois theory, even algebraic topology – he is probably best known today for his views on the philosophy of mathematics, especially for his philosophical differences with Weierstrass and Cantor. There is irony in this, because in one of his few pronouncements on the philosophy of mathematics he stated that his goal was always to *avoid* philo-

Leopold Kronecker

sophical issues and to base his mathematics on formulas and algorithms that were unexceptionable. He dedicated himself to proving that mathematics can be done in this way, but his colleagues and later generations of mathematicians have not valued his achievements in this direction, but instead have come to see him, in Hilbert's term, as a *Verbotsdiktator* (see [11], pp. 159 and 161). The oft-quoted statement of Poincaré to the effect that Kronecker's successes were achieved by ignoring his philosophy is a quotation out of context ([7]) that does not even represent Poincaré's true opinion, much less a correct statement about Kronecker's work; his papers are in fact permeated by his algorithmic philosophy, and have much in common with present day work on computer algebra. In this respect, Kronecker's work is more modern than the "modern algebra" of the period between his time and ours.

Kronecker's career was as unconventional as his mathematics. He was born to a prosperous Jewish family in Liegnitz (see [9], cf. [12], [1], [3]). His father, a well-educated and cultured merchant who wanted the best classical education for his son, first hired a private tutor for him and then sent him to the *Gymnasium* in Liegnitz. His studies were very broad, including the classical languages, philosophy, history, literature, and theology, but already at this early stage of his career his commitment to mathematics was won by his mathematics teacher, and later lifelong friend, E.E. Kummer, who was at the time a young teacher at the *Gymnasium*.

He entered the University of Berlin in 1841, at the age of 17. After four years – part of them spent at the universities of Bonn and Breslau – he received his doctorate for a thesis in algebraic number theory that was extremely highly regarded by Dirichlet, his principal teacher in Berlin.

But after he received his doctorate in 1845 he in no way embarked on an academic career. Instead, heeding his father's wish, he returned to Silesia and undertook the management of a large estate owned by his family. When his uncle died the following year, Kronecker switched from agriculture to banking and took charge of the liquidation of the uncle's banking business. He appears to have had as great a talent for business as for mathematics. His wealth may, of course, have been mostly inherited, but it is certain that by the time he wound up his business affairs and moved to Berlin in 1855, he was financially independent. He had married the daughter of his banker uncle in 1848, and they eventually had six children. They lived in a grand house in Berlin, travelled frequently within Germany and abroad, and enjoyed all the comforts and advantages of wealth. When Kummer retired in 1883, Kronecker succeeded him as a professor at the university, but this appears to be the only job Kronecker ever held.

Although he published no mathematics between 1845 and 1853, his correspondence with Kummer and Dirichlet shows he was still quite active mathematically. The 1853 paper with which he ended his silence contained the statement of what is now called the Kronecker-Weber theorem (Heinrich Weber was 10 years old at the time): Every abelian extension of the field of rational numbers is a subfield of a cyclotomic field. In fact, this paper even ends with a hint of what was later called Kronecker's *Jugendtraum* theorem regarding the abelian extensions of imaginary quadratic fields. This is the first of many instances of Kronecker letting it be known that he was in possession of important – even dazzling – results without publishing them or backing them up with proofs.

Another such instance is his generalization of Kummer's "ideal prime factors" to arbitrary algebraic number fields. Characteristically, Kummer had developed the theory only in the case he needed for his study of reciprocity laws, namely, the cyclotomic case. Dedekind recognized early on the desirability of a generalization that would apply to arbitrary algebraic number fields, but the difficulties were great. In 1858, well before Dedekind was engaged with the problem, Kronecker wrote a paper that contained, he later said, a complete generalization, but he never published it. Kummer announced in a paper published in 1859 that Kronecker would "soon" publish a complete generalization, but in fact it was not until 1882 that Kronecker sketched such a theory in his Kummer *Festschrift*; there he refers to the 1858 paper but makes no apology for not having published it. (Meanwhile, Dedekind had worked out and published his own epoch-making – and profoundly different – generalization of Kummer's theory in 1871.)

Camille Jordan, in the introduction to his treatise on substitutions and algebraic equations of 1870, confessed that he would be unable to cover "these beautiful theorems" of Kronecker in the theory of algebraic equations "which are at the present time the envy and the despair of geometers." The despair has only increased. His works, difficult to read at the time, are even more difficult today, now that the culture in which they were embedded is largely lost. An indication of the difficulty of Kronecker's work is given by a lengthy note by Helmut Hasse in the last volume of Kronecker's collected works. This note concerns the statement and proof of Kronecker's *Jugendtraum*. As far as the statement is concerned, Hasse refutes the interpretation Hilbert and Fueter had made – an interpretation he himself had propagated in his *Klassenkörperbericht*. The statement Hilbert and Fueter attributed to Kronecker was demonstrably false, but Hasse shows that it is certainly not the one Kronecker intended. As for the proof, Hasse believed that the series of major articles on elliptic functions that Kronecker began publishing in 1883 and was continuing to work on at the time of his death, would have concluded with a proof of the *Jugendtraum* theorem.

Another difficult early work of Kronecker was his class number relations – a type of useful, subtle, and baffling relations among the class numbers of binary quadratic forms for different determinants that he derived from the theory of elliptic functions with complex multiplication.

Obscure though it might have been, the importance and the depth of Kronecker's work was beyond doubt. In 1861 he was elected to the Berlin Academy. This honor carried with it the "right" to give courses of lectures at the university, and Kronecker regularly made use of this right during the next 22 years, before he became a professor.

Kronecker had the opportunity to become a professor much sooner, and in a spectacular way. When Riemann died in 1866, the chair in Göttingen that he left vacant was offered to Kronecker. As Kronecker said in a letter to Lifschitz ([15], p. 170) some years later, this chair had been "raised by the names of Gauss, Dirichlet, and Riemann almost to a mathematical *throne*." Nonetheless, after hard deliberation, he decided that his life in Berlin, and especially his connection with his colleagues Kummer, Weierstrass, and Borchardt, and his activities at the Berlin Academy and the university, were more attractive to him than the academic life in Göttingen.

The collaboration of Kummer, Kronecker, and Weierstrass in the first decades of their years together in Berlin were filled with mutual respect and with cross-fertilization of ideas (see [13], p. 498). Even as teachers they worked closely together. Georg Cantor was formally a student of Kummer, receiving his doctorate in 1867, but his ties to Kronecker and Weierstrass were strong during his student days and beyond. The same was true of other Berlin students.

In the years of Kronecker's maturity, an immense change overcame mathematics: The classical rejection of completed infinites, which had been maintained by Gauss and most other mathematicians of the earlier period, was overthrown. The revolution was not that the statement "an arbitrary sequence of real numbers in the interval $0 \leq x \leq 1$ has a convergent subsequence" came to be accepted as *true*, but that it came to be accepted as *meaningful*. Kronecker contended that the acceptance of completed infinites was unnecessary and he devoted himself to doing his mathematics without it. This view not only put a barrier between him and younger mathematicians like Cantor, Schwarz, Dedekind, and Hilbert, it poisoned his relations with his older colleague Weierstrass. In an oft-quoted letter (24 March 1885, published by Mittag-Leffler in the proceedings of the ICM in Paris, 1900, and by others since) to Sofia Kovalevskaia, Weierstrass complained that Kronecker was making "a direct appeal to the younger generation to forsake those who have been their leaders up until now and to follow him as the apostle of a new doctrine." If Weierstrass ever confronted Kronecker with this complaint, it is likely that Kronecker would have replied that in reality it was Weierstrass's doctrine of completed infinites that was new, and that students should hear competing views and be encouraged to think for themselves, not to follow leaders. Later generations, enthusiastic in their support of the new mathematics, and especially in their support of Cantor and Weierstrass, have depicted Kronecker as a harsh and dogmatic critic of the new views, but there appears to be no documentary evidence for this depiction of Kronecker; there is definitely no evidence he made personal attacks on those with whom he differed.

Kronecker's wife, Fanny, died on August 23, 1891. It is said he was so stricken with grief that he lost the wish to live. He died on December 29 of the same year, of bronchitis. The volume of his publications, which was far greater in the 1880s than it had been earlier, was maintained nearly to the end of his life, in quality as well as quantity. Still, many of his ideas were never published in finished form. His scientific heir, Kurt Hensel, received, in his own words, "a rich and carefully preserved scientific *Nachlass,*" including many works that were "fully or partly prepared" for publication. Alas, Hensel never fulfilled his promise to publish material from this *Nachlass,* and it was lost forever in an explosion in 1945 [8]. Kronecker's works remain, as Jordan said, objects of envy and despair.

References

[1] *Biermann, K.-R.*, Die Mathematik und ihre Dozenten an der Berliner Universität 1810–1933, Akademie Verlag, Berlin, 1988.

[2] *Biermann, K.-R.*, E.E. Kummer, in Dictionary of Scientific Biography.

[3] *Biermann, K.-R.*, Leopold Kronecker, in Dictionary of Scientific Biography.

[4] *Edwards, H.M.*, The background of Kummer's proof of Fermat's Last Theorem for regular primes, Arch. Hist. Exact Sci. **14** (1975), 219–236.

[5] *Edwards, H.M.*, Postscript to "The background of Kummer's proof of Fermat's Last Theorem for regular primes", Arch. Hist. Exact Sci. **17** (1977), 381–394.

[6] *Edwards, H.M.*, Fermat's Last Theorem, A genetic introduction to algebraic number theory, Graduate Texts in Math. 50, Springer-Verlag, Berlin, 1977.

[7] *Edwards, H.M.*, Kronecker on the Foundations of Mathematics, in: Jaakko Hintikka (ed.), From Dedekind to Goedel, Kluwer, Dordrecht, 1995, 45–52.

[8] *Edwards, H.M.*, On the Kronecker Nachlass, Hist. Math. **5** (1978), 419–429.

[9] *Frobenius, G.F.*, Gedächtnisrede auf Leopold Kronecker, Abhandlungen der Königlich Preussischen Akademie der Wissenschaften zu Berlin, 1893, also in: Collected works, vol. 3, 705–724.

[10] *Hensel, K.*, Gedächtnisrede auf Ernst Eduard Kummer, in: Festschrift zur Feier des 100. Geburtstages Eduard Kummers, Teubner, Leipzig und Berlin, 1910, also in: Weil, A. (ed.), E.E. Kummer, Collected Papers , vol. I, Springer-Verlag, Berlin, 1975, 33–69.

[11] *Hilbert, D.*, Neubegründung der Mathematik, Erste Mitteilung, Abh. Math. Sem. Hamburg **1** (1922), 157–177, also in: Gesammelte Abhandlungen, vol. III, Springer-Verlag, Berlin, 1935, 157–177.

[12] *Kneser, A.*, Leopold Kronecker, Rede, gehalten bei der Hundertjahrfeier seines Geburtstages in der Berliner Mathematischen Gesellschaft am 19. Dezember 1923, Jber. dt. Math.-Verein. **33** (1925), 210–228.

[13] *Kronecker, L.*, Auszug aus einem Briefe von L. Kronecker an G. Cantor, in: Hensel, K. (ed.), Leopold Kroneckers Werke, vol. V, Leipzig, 1930.

[14] *Lampe, E.*, Nachruf für Ernst Eduard Kummer, Jber. dt. Math.-Verein. **3** (1893), 13–28, also in: Weil, A. (ed.), E.E. Kummer, Collected Papers, vol. I, Springer-Verlag, Berlin, 1975, 15–30.

[15] *Lifschitz, R.*, Briefwechsel mit Cantor, Dedekind, Helmholtz, Kronecker, Weierstrass und anderen, bearbeitet von W. Scharlau, Dokumente zur Geschichte der Mathematik, Band 2, DMV, Vieweg Verlag, 1986.

[16] *Meth, B.*, Ernst Eduard Kummer als Lehrer, Mathematischer Verein Berlin, 1910, unpublished.

Weierstrass and some members of his circle: Kovalevskaia, Fuchs, Schwarz, Schottky

Reinhard Bölling

Weierstrass

Karl Weierstrass (1815–1897) belongs to the outstanding mathematicians who have worked in Berlin. After 13 years as a *Gymnasium* teacher at remote locations in Prussia far away from the centers of mathematical research, Weierstrass came to Berlin in 1856 at the age of 41 (as professor extraordinarius; he was promoted to ordinarius in 1864). This advancement in his career came after he solved one of the era's challenging problems: the Jacobi inversion problem for hyperelliptic integrals, first formulated in 1832. Weierstrass published a preliminary version of his solution in 1854 in *Crelle's Journal*, and his results served as the starting point for the emergence of the theory of Abelian functions. Weierstrass's paper created a sensation, and from one day to the next his name became well known in mathematical circles. The aged A. von Humboldt (1769–1859) along with E. E. Kummer (1810–1893) emphatically supported his appointment, and in the years that followed he became one of the leading mathematicians of his time. Together with his colleagues Kummer and L. Kronecker (1823–1891), Weierstrass ensured Berlin's reputation as a world-class mathematical center in the second half of the 19th century.

Every student of mathematics encounters the name of Karl Weierstrass in their first semesters at the university, since a lecture course in analysis would be unthinkable without treating fundamental notions that Weierstrass invented or used. Nearly every mathematician today knows the Bolzano-Weierstrass theorem, Weierstrass's criterion for uniform convergence or Weierstrass's theorem on uniformly convergent series of analytic functions. With his name are connected such significant results as the preparation theorem, the theorem on the approximation of functions, the theorem on double series, the infinite product theorem, the Casorati-Weierstrass theorem, and the Weierstrass-Erdmann corner conditions. For nearly two centuries before him, many excellent mathematicians had striven to develop an exact foundation for analysis, struggling, for instance, with the problem of "infinitely small quantities". Weierstrass gave the subject the general form in which it is known to mathematicians today. He carried out his proofs with the greatest care and attention to detail, and this "Weierstrassian rigour" soon became legendary. Exemplary, in this regard, was his treatment of infinite series. Weierstrass realized the fundamental importance of the concept of uniform convergence and he was the first in using it systematically for the foundation of analysis (the concept itself was hit upon independently by others, too).

Karl Weierstrass

Together with A.L. Cauchy (1789–1857) and B. Riemann (1826–1866), Weierstrass belongs to the founders of the theory of functions of complex variables. Central in Weierstrass's function theory is the notion of the power series and its analytic continuation. He contributed fundamental results to the variational calculus and differential geometry. His understanding of rigour led him to avoid geometrical arguments, and thus it was he who criticized one pillar of Riemann's function theory, namely the so-called Dirichlet principle. The nascent theory of elliptic functions, as developed by A.-M. Legendre (1752–1833), N.H. Abel (1802–1829) and C.G.J. Jacobi (1804–1851), formed a focal point of research in mathematics of the 19th century. Under Weierstrass this field experienced a major reconstruction (based on the Weierstrass $\wp$-function). Beyond this achievement, Weierstrass aimed to establish an exact foundation for the theory of Abelian functions, a vast domain that contains the elliptic functions as its most important special case. This goal stood at the center of Weierstrass's scientific efforts throughout his career.

The exact foundation of analysis required an exact foundation of the concept of a real number. Weierstrass presented his theory of real numbers, for the first time, during the winter semester 1863/64 in his lectures on analytic functions. He himself never published his approach (as is true of many other of his results). The basic principle of Weierstrass's theory consists in the formation of (infinite) sums (one has to formalize the idea that every real number is a sum of rational numbers). Quite different approaches were given by R. Dedekind (1831–

1916) (formation of "cuts"), Ch. Méray (1835–1911) and G. Cantor (1845–1918) (fundamental sequences), where Cantor's theory proved to be the most capable of generalization.

His example from 1872 of a continuous, but nowhere differentiable real function served, like all his contributions to the conceptual foundations of analysis, his overall goal mentioned above. Soon a substantial part of the mathematical community realized the neccessity of strong foundations for the basic concepts of analysis. Even so, twenty years later such a function represented in the eyes of Ch. Hermite (1822–1901) a "deplorable wound".

Some basic principles

Weierstrass expressed his views on the principles of scientific research in a letter to Kovalevskaia of 1 January 1875: "What I insist on in scientific research is unity of method, the systematic pursuit of a definite plan, and the appropriate following through of the details and – that the work carries the mark of independent research."

Lectures

Weierstrass achieved great effectiveness through his lectures. These became a drawing-point for students as well as mature mathematicians from both Germany and abroad. Much of the material presented by Weierstrass had not been published, including most recent results of his own research (as, for instance, the above-mentioned theory of real numbers or the refoundation of the theory of elliptic functions which he lectured on for the first time in the winter semester 1862/63). In the 30 years of his teaching at the University of Berlin, Weierstrass presented his lecture courses in a fairly regular cycle: Introduction to the theory of analytic functions (including the theory of real numbers) – Elliptic functions – Abelian functions – Application of the elliptic functions – Calculus of variations. The number of his auditors was impressive (and, by the way, Kummer's too). Even the demanding lecture on Abelian functions was attended at times by 200 registered students. It was to a great extent the lecture notes (a considerable number of which have survived; many of them are detailed and carefully written) that essentially supported the further spreading of Weierstrass's mathematics throughout Europe. The list of later renowned mathematicians who had attended lectures of Weierstrass is impressive. Among them were (without any intention for completeness): O. Bolza (1857–1942), A. Brill (1842–1935), H. Bruns (1848–1919), H. Burkhardt (1861–1914), G. Cantor, F. Engel (1861–1941), R. Fricke (1861–1930), G. Frobenius (1849–1917), L. Fuchs (1833–1902), L. Gegenbauer (1849–1903), A. Gutzmer (1860–1924), K. Hensel (1861–1941), O. Hölder (1859–1937), A. Hurwitz (1859–1919), W. Killing (1847–1923), A. Kneser (1862–1930), L. Königsberger (1837–1921), S. Kovalevskaia (1850–1891), M. Lerch (1860–1922), J. Lüroth (1844–1910), Hj. Mellin (1854–1933), F. Mertens (1840–1927), H. Minkowski (1864–1909), G. Mittag-Leffler (1846–1927), H. v. Mangoldt (1854–1925), E. Netto (1846–1919),

L. Pochhammer (1841–1920), A. Pringsheim (1850–1941), F. Rudio (1856–1929), C. Runge (1856–1927), L. Schlesinger (1864–1933), A. Schönflies (1853–1928), F. Schottky (1851–1935), F. Schur (1856–1932), H.A. Schwarz (1843–1921), P. Stäckel (1862–1919), L. Stickelberger (1850–1936), O. Stolz (1842–1905). We add that the physicists L. Boltzmann (1844–1906), M. Planck (1858–1947) and the philosopher E. Husserl (1859–1938) had attended lectures of Weierstrass, too. Weierstrass drew up the first expertises for 28 dissertations. Apart from Kummer (with 39 first expertises) no academic teacher of mathematics at the University of Berlin ever (at least until 1936) achieved such a number of doctoral students.

It is a fact that Weierstrass at the peak of his fame was in higher regard than the versatile genius Riemann. In the words of A. Kneser, Weierstrass was the "unquestioned ruler of the whole enterprise". Not everybody was willing to accept this authority and it was, for instance, the reason why F. Klein (1849–1925) and S. Lie (1842–1899) did not attend a lecture of Weierstrass during their time of study in Berlin. To be sure, Klein regretted it later on.

The Mathematical Seminar

One of the important events for the organisation of the mathematical life of the University of Berlin was the foundation of the first seminar devoted exclusively to mathematics. An initiative of Kummer and Weierstrass, the Berlin seminar began in 1861. The number of participants was limited to twelve, and admission required submitting a paper or passing an examination. One effect of the seminar was that talented students could be recognized and supported early on. Among the participants one finds numerous scholars later to become university professors. To support the seminar, a library was established, and participants could be awarded a prize. Kummer and Weierstrass led the seminar for almost two decades. At the same time (also in 1861) an association, the *Berliner Mathematischer Verein*, was founded, with the aim of providing a forum for all the other students of mathematics in Berlin. An average of 50 members belonged to this association.

Weierstrass and Cantor

Cantor's publications on set-theory (since 1872) were met by his contemporaries mostly with rejection or scepticism. Weierstrass was open-minded towards these unaccustomed ideas, although he only rarely stated his point of view about Cantor or the latter's set-theoretical research. Weierstrass occasionally used notions or results from Cantor's work. We know that he sometimes encouraged Cantor to publish his findings. Time and again, personal meetings took place between the two. Also as far as his own set-theoretical work was concerned, Cantor appreciated (and searched for) the judgement of the "brilliant great function theorist", as he called him once. But occasionally Cantor was also critical of Weierstrass.

Controversy with Kronecker

Not later than 1870, Weierstrass and Kronecker adopted opposing positions concerning the foundation of analysis. More precisely, they disagreed on the treatment of the infinite in mathematics. Kronecker did not accept a pure existence statement such as is given, for instance, by the Bolzano–Weierstrass theorem. Kronecker took the basic view that to the foundation of analysis only constructive procedures are admissible that in every single case require only a finite number of steps. For instance, the introduction of the notion of irreducibility of a polynomial is only justified if, at the same time, a procedure is stated which allows to decide in a finite number of steps whether a given polynomial is irreducible or not. For this reason, it is also quite clear that Kronecker could not accept Cantor's set-theoretical work.

The relationship between Weierstrass and Kronecker who once were friends eventually grew quite strained. In the 1880's, the controversy between them became even more acute and led, finally, to a break between them. In public, however, they did not oppose one another. Kronecker spoke of the "so-called analysis". For Weierstrass the strain was so great that he even had the intention, a few weeks before his 70th birthday, to leave Berlin and settle in Switzerland.

End of an era

With the death of Weierstrass in 1897 the last representative of the golden age of mathematics in Berlin was gone. An era of mathematics had come to an end, and over the next two decades Berlin was overtaken by Göttingen which, under Klein and D. Hilbert (1862–1943), emerged as the leading mathematical center in Germany.

Kovalevskaia

Among the pupils of Weierstrass, the Russian Sofia Kovalevskaia (1850–1891) takes a special place. A great number of publications have been devoted to the life and work of this fascinating personality.

In the autumn of 1870 she had come from Heidelberg to Berlin in order to continue her studies with Weierstrass. At the age of 18 she married V.O. Kovalevski (1842–1883), who later became a well-known paleontologist. It was a marriage on paper only designed to allow her to study abroad. But even Weierstrass did not succeed in receiving permission for the young Russian woman to attend university lectures. (Prussia first officially allowed women to attend university only in 1908.) Convinced of her exceptional talent, Weierstrass offered her private tuition (free of charge). Thus, she became his pupil. Kovalevskaia remained in Berlin until the summer of 1874, and a relationship grew between them which is difficult to find anywhere else in the history of science. Only her untimely death set an end to this friendship.

In August 1874 she received the doctoral degree from the University of Göttingen, the first woman anywhere ever to attain this status in mathematical sciences. One of her three manuscripts submitted to Göttingen contained the theorem guaranteeing the existence of an analytic solution to a certain type of partial differential equation with analytic coefficients (equations of normal form) in the

Sofia Kovalevskaia

case of analytic initial values; this result is now commonly known as the Cauchy-Kovalevskaia theorem. In 1884 she was appointed professor for higher analysis at the then still young University of Stockholm. For the first time, a woman held a chair in mathematics in an unrestricted sense. (In 1750 M.G. Agnesi (1718–1799) had been appointed as professor but gave her lectures only for two years and then withdrew more and more from scientific research.) A peak of Kovalevskaia's scientific career occurred when she received the prestigious Prix Bordin of the French Academy of Sciences in 1888 for her results on the rotation of a rigid body about a fixed point. The solution of the corresponding differential equation derived by L. Euler (1707–1783) had been obtained in two special cases studied by Euler and by J.L. Lagrange (1736–1813). Kovalevskaia found a third case that could be solved by means of certain Abelian functions. This case is in so far the last one possible under the condition that the solutions should be single-valued and meromorphic functions of time. She had become well-versed with the theory of these functions due to her studies under Weierstrass.

Kovalevskaia also worked as a writer. Her greatest success were the reminiscences of her childhood years which first appeared in 1889 in Stockholm. They were translated into several languages and are still reprinted to this day.

Lazarus Fuchs

Fuchs: Between Berlin's Classical Era and Its First Post-Classical Period

Lazarus Fuchs (1833–1902) studied in Berlin where he attended the lectures of Kummer and Weierstrass, among others. He was awarded a doctorate in 1858 (examiners of his dissertation: Kummer, M. Ohm (1792–1872; brother of the physicist G.S. Ohm)). In 1865 he became a *Privatdozent* (lecturer), and the following year he was appointed *Extraordinarius* (associate professor) at the University of Berlin, where he taught until the winter semester 1868/69. Subsequently, he received appointments in Greifswald, Göttingen (1874) and Heidelberg (1875). In 1884 he returned to Berlin as the successor to Kummer. Thus, Fuchs is a representative of both Berlin's classical and its first post-classical era. His personality has been described as indecisive, timid, but at the same time humorous and full of kindness.

Fuchs enriched the theory of linear differential equations with fundamental results. He discussed problems of the following kind: What conditions must be placed on the coefficients of a differential equation so that all solutions have prescribed properties (e.g. to be regular or algebraic). This led him (1865, 1866) to introduce an important class of linear differential equations (and systems) in the complex domain with analytic coefficients, a class which today bears his name (Fuchsian equations, equations of the Fuchsian class). For instance, the hypergeometric equation is of the Fuchsian class, and, moreover, it was possible to characterize the class of such equations as Fuchs observed (1865, 1866). He also succeeded in characterizing those linear differential equations the solutions of which have no essential singularity in the extended complex plane. Fuchs later also studied non-linear differential equations and movable singularities.

Fuchs's study (1876; together with Hermite) of elliptic integrals as a function of a parameter marks an important step towards the theory of modular functions (Klein, Dedekind). In a series of papers (1880–1881) Fuchs studied functions obtained by inverting the integrals of solutions to a second-order linear ordinary differential equation in a manner generalizing Jacobi's inversion problem. As a special case, the function inverse to the quotient of two independent solutions occurs, and this function proved to be invariant under a certain group of transformations. The study of the nature of this inverse function was the starting point for the investigations of H. Poincaré (1854–1912), and led him finally to introduce the concept of the Fuchsian group. The theory of Fuchsian groups came to form a basic element in the magnificent theory of automorphic functions created by Poincaré and Klein (in honor of Fuchs, Poincaré used the term "fonctions Fuchsiennes") with all of its fundamental applications to the theory of uniformization.

Fuchs taught at the University of Berlin until his death in 1902.

Schwarz: A Figure of Berlin's First Post-Classical Era

Hermann Amandus Schwarz (1843–1921; son-in-law of Kummer) came to Berlin as Weierstrass's successor in 1892. This marks the beginning of the first period of Berlin's post-classical era. Schwarz's collected papers had already been published in two volumes in 1890. Actually, Schwarz's important mathematical works date from the time before his appointment to Berlin. L. Bieberbach (1886–1982) characterized this by the witty remark that Schwarz retired to his Berlin chair.

Schwarz had spent all of his student years in Berlin. He obtained his doctorate in 1864 (examiner of his dissertation: Kummer). In 1867 he became a *Privatdozent* (lecturer) at the University of Halle. In 1869 he was appointed professor at the *Eidgenössische Technische Hochschule* in Zurich. From 1875 on he taught at the University of Göttingen.

Schwarz was deeply influenced by Weierstrass. From their correspondence one finds that Schwarz addressed his teacher often with an accuracy going down to the last detail, sometimes almost timidly. Schwarz's demeanor has been described as naive, dramatic, coarse. In spite of giving the impression of self-confidence, he was, in fact, rather insecure and, besides, not efficient in business matters.

Unlike Weierstrass, Schwarz had an inclination to the geometrical approach. At times, ideas coming from geometrical considerations were translated into the language of analysis. Thus, Riemann's kind of thoughts could become fruitful in combination with Weierstrass's mathematics. The idea of the universal covering surface, which goes back to Schwarz, is an example of this. By means of this concept Poincaré succeeded (1883) in proving the uniformization of analytic functions of one variable in a number of cases. Schwarz made important contributions to the theory of minimal surfaces. The discovery of what has come to be known as the Schwarz minimal surface was published in 1865 (a minimal surface whose

Hermann Amandus Schwarz

boundary consists of four edges of a regular tetrahedron). He developed a method of successive approximation to solve differential equations. Schwarz invented the important alternating method (1870) – one of the general methods for finding a solution to the Dirichlet problem (the problem of constructing a harmonic function from its boundary values). This allowed him to prove Riemann's mapping theorem for domains, the boundaries of which satisfy certain restrictions, a result that was in those days considered as one of the most difficult in all of analysis. Schwarz's famous mapping theorem also arose in this research, a result that has found its fixed place in function theory since then. Riemann had based his proof on the so-called Dirichlet principle (in his 1851 dissertation), which remained unjustified after the criticism of Weierstrass in 1870 (in fact, the principle had been discussed already earlier as, for instance, the work of Schwarz just mentioned shows). Thus, Schwarz's methods were (together with those of C. Neumann (1832–1925)) for decades the only access to Riemann's theorem. Finally, Hilbert succeeded in 1901 in rehabilitating the principle, putting it at last on a rigorous basis.

Besides his achievements in the field of conformal mapping (e.g. the reflection principle (1869)), Schwarz made significant contributions to the calculus of variations, and to potential theory. His study of Gauss's hypergeometric series led Schwarz to an analytic function (now called the Schwarz function) by means of which a conformal mapping of a triangle bounded by arcs of circles onto the upper half-plane (or unit disc) can be realized. (His function was a further early example for an automorphic function.) In the special case of a rectilinear triangle these functions are known as Schwarz triangle functions that can be represented as the quotient of two independent solutions of a hypergeometric equation. For the study of boundary value problems in the theory of analytic functions, the Schwarz integral (and its generalizations) is very important. Several other objects bear his name today (e.g. the Schwarz differential, Schwarz equation, Schwarz kernel, Schwarz surface, Schwarz derivative).

Schwarz taught at the University of Berlin until 1918.

Schottky: A Figure of Berlin's Second Post-Classical Era

Friedrich Schottky (1851–1935) studied in Breslau and Berlin. He was awarded the doctoral degree in 1875 (examiners of his dissertation: Weierstrass, Kummer). In 1878, he became a *Privatdozent* (lecturer) at the University of Berlin. He later received appointments in Zurich (1882) and Marburg (1892). In 1902 he returned to Berlin as Fuchs's successor. Schottky was a quiet, inconspicuous researcher. While studying Schottky's dissertation, Weierstrass wrote to Kovalevskaia in a letter of 7 May 1875 that this work (on conformal mappings of multiply connected domains, cf. below) is one of the best he ever examined, and he went on to say: "The author is of a clumsy appearance, unprepossessing, a dreamer, but, if I'm not completely wrong, he possesses an important mathematical talent. On Christmas Eve he was suddenly arrested and led away to the barracks to serve a 3-year term as a common soldier, for 'he had forgotten' to register in time for a 1-year term as a volunteer (*Freiwilliger*) (as every student does). Fortunately, he proved to be so useless as a soldier that he was discharged as unsuitable after 6 weeks. Thus, he could return to his dissertation. He then signed up for the examination without [presenting the requisite] certificates and without knowing anything about the neccessary formalities. As rector I had to cancel his name from the register because neither had he attended lectures nor were his whereabouts in Berlin known." (Weierstrass had been elected rector for the academic year 1873/74.)

Schottky's teaching qualities were not impressive ("As a teacher he is unsuitable", Kummer once said). Perhaps, for this reason, he almost never taught beginners' classes in Berlin.

In 1875, Schottky discovered a type of function which was later intensively studied, in full generality, by Poincaré and Klein (automorphic functions). He was the first to study systematically conformal mappings of multiply connected domains. In the case of a domain bounded by p closed curves ($p \geq 2$), Schottky discovered that there exist $3p - 3$ real constants that characterize the conformal class to which this region belongs. These are now called the *moduli* of the class. Schottky is best known for his significant contributions to the theory of Abelian functions. The problem of determining which principally-polarized Abelian varieties (of dimension g) are Jacobian varieties (of an algebraic curve of genus g) is known as the Schottky problem (Riemann had already posed this question). In 1888, Schottky gave a first result in the case $g = 4$ (for $g \leq 3$ the varieties coin-

Friedrich Schottky

cide). He later proved (1904) a generalization of Picard's 'Great' Theorem, which is now a classical result in the theory of functions of a complex variable that bears his name.

Schottky taught at the University of Berlin until 1926.

Acknowledgement

The author would like to thank M. Korey and D.E. Rowe for their assistance in correcting parts of his English text.

Selected references

[1] *Biermann, K.-R.*, Die Mathematik und ihre Dozenten an der Berliner Universität 1810–1933, Akademie Verlag, Berlin, 1988.

[2] *Bölling, R.*, Briefwechsel zwischen Karl Weierstrass und Sofja Kowalewskaja, Herausgegeben, eingeleitet und kommentiert von R. Bölling, Akademie Verlag, Berlin, 1993.

[3] *Bölling, R.*, Karl Weierstrass – Stationen eines Lebens, Jber. dt. Math.-Verein. **96** (1994), 56–75.

[4] *Cooke, R.*, The Mathematics of Sonya Kovalevskaya, Springer-Verlag, New York, 1984.

[5] *Gray, J.*, Linear differential equations and group theory from Riemann to Poincaré, Birkhäuser, Boston, Basel, Stuttgart, 1986.

[6] *Hibner Koblitz, A.*, A convergence of lives, Sofia Kovalevskaia: scientist, writer, revolutionary, Birkhäuser, Boston, Basel, Stuttgart, 1983.

[7] *Kolmogorov, A.N., Yushkevich, A.P. (eds.)*, Mathematics of the 19th century, Geometry, Analytic function theory, Engl. trans. R. Cooke, Birkhäuser, Basel, Boston, Berlin, 1996.

Frobenius, Schur, and the Berlin Algebraic Tradition

Ralf Haubrich

Ferdinand Georg Frobenius (1849–1917) and Issai Schur (1875–1941) ranked among the most distinguished representatives of what may be called the Berlin algebraic tradition. The tradition originated with Frobenius's teachers Weierstrass and Kronecker, was carried on by Frobenius, and handed down from Frobenius to Schur who was its last prominent "pure" member. Weierstrass, Kronecker, Frobenius, and Schur were full professors of mathematics at the University of Berlin and probably the most influential proponents of this tradition. They shared common views on what kind of problems and theories are worth pursuing, what statements are important, which methods and what mathematical languages are to be used, etc.; in particular, they preferred a formal, algebraic approach to mathematics (see [11], ch. 1, and [38]).

Frobenius was both personally and scientifically closely bound to Berlin (the following biographical remarks are drawn from [2], pp. 116–172 passim, pp. 287–290). He was born there, studied there (with the exception of one semester in Göttingen), and in 1870 he received his doctorate in Berlin with a dissertation under the direction of Weierstrass. In 1874, after some years of teaching at the high school level – likewise in his native town – he became associate professor at the University of Berlin. However, he spent the years from 1875 to 1892 in Zürich where he was a full professor at the *Polytechnikum.* But following this, he returned to Berlin to take over Kronecker's chair. Weierstrass regarded Frobenius as one of his most gifted students. It was largely due to his influence that the professorships in Berlin (1874 and 1892) were offered to Frobenius.

1892 is seen as the beginning of a new period in the history of Berlin mathematics because two of a total of three chairs were refilled (those of Kronecker and Weierstrass). However, the continuity had never been endangered. All professors of mathematics between 1892 and 1917 (when Frobenius died) had studied in Berlin. They all felt deeply obliged to carry on the Prussian neo-humanistic tradition of university research and teaching as they themselves had experienced it as students (for the Prussian neo-humanistic tradition see [2], pp. 186–189, [52], and [57], pp. 1444–1446). This is especially true for Frobenius. He considered himself to be a scholar whose duty it was to contribute to the knowledge of pure mathematics. Applied mathematics, in his opinion, belonged in the technical colleges.

From 1892 to 1917, Frobenius was the dominant personality of Berlin mathematics, occasionally choleric, quarrelsome, and given to invectives. He felt a particular aversion to Göttingen mathematics. The reason for this was, above all, Klein's work in scientific policy and organization. It was partly due to the influence of Klein that around 1890 the institutional and educational system of German

Ferdinand Georg Frobenius

mathematics had begun to change (see [57], pp. 1448–1454). This transformation resulted in a departure from the neo-humanistic ideal. Frobenius condemned this departure but was powerless to do anything about it. He was only able to exercise influence in Berlin. It was to a large extent Frobenius who removed Berlin mathematics from the development in Germany and who, as a consequence, led Berlin into comparative isolation.

Klein's orientation was doubly unbearable for Frobenius because the years 1892–1917 marked a period of temporary decline of Berlin mathematics (see [2], pp. 157–159, 171) whereas Göttingen mathematics flourished (see [54] and [99]). Berlin suffered a considerable loss of both attractiveness and influence. Symptomatic of this is the fact that in the twenty-five years of his activity Frobenius supervised only ten doctoral candidates. One of the reasons for the decline was the increasing isolation of Berlin mathematics. Another reason, although less important, can be found in Frobenius's teaching activities. His courses – usually on number theory, algebra, determinants, and analytic geometry – were solid, comprehensive, and well-expressed but not very stimulating. And more important, he was seldom personally involved in the education of students (see [12], pp. 118, 119, and [96]).

However, as a researcher Frobenius enjoyed an excellent reputation. He published articles in an uncommonly large number of different disciplines of pure mathematics. (1) His first larger field of work (from 1873 to 1879) involved differ-

Issai Schur

ential equations (see [30], pp. 79–87, 145, 146; for a survey see [55]). He succeeded in rederiving important parts of Fuchs's fundamental work in a new and simpler way. In doing so he introduced methods and concepts which have since become customary in the treatment of differential equations.

Frobenius's versatility first became apparent in the late 1870s: (2) Mostly together with Stickelberger he published a number of articles on elliptic functions. (3) In 1880 the first in a series of articles on theta functions appeared. This was Frobenius's main area of interest in the 1880s. (4) Already in the late seventies he was familiar with Kronecker's number-theoretical work and, most noteworthy, Dedekind's then mainly unstudied algebraic number theory (see [9], pp. 414–419, and [10], pp. 277, 278). Of his results he published very little; the most important contribution is the so-called Frobenius density theorem (in [16]).

(5) Motivated by number-theoretical notions and results, Frobenius and Stickelberger published a comprehensive paper on finite abelian groups [14]. It contains a proof of what now is called the basic theorem of those groups. This paper was the definitive publication on the subject for many years (see [47], p. 324, and [101], pp. 234, 235).

(6) His most influential papers prior to the 1890s, however, dealt with the theory of bilinear forms as developed by Weierstrass and Kronecker (for a detailed account see [34], pp. 569, 570, [35], pp. 96–101, and [36], pp. 147–155). In 1878 and 1879 he combined this theory with the symbolical calculus of linear substitutions

under composition and thus created linear algebra as we now conceive it. Later he applied this theory to other disciplines whereby from the 1890s onward he more and more explicitly used the language of matrices. (7) Around 1910 he published papers on positive and non-negative matrices which included the introduction of the concept of irreducibility for a matrix. Frobenius's results have remained fundamental to the theory of non-negative matrices (see [56]).

(8) The bulk of Frobenius's memoirs on pure group theory appeared from 1893 onwards. He was mainly concerned with the relation between the arithmetic properties of the order of a finite group and its structure: he generalised Sylow's theorems and gave new proofs of the original ones (see [8], pp. 50–58, [48], pp. 78–81, and [100], in particular pp. 282–288), and he studied solvable and simple groups (see [97], pp. 343–347). Between approximately 1880 and 1910 group theory, as we now conceive it, was shaped. Although Frobenius was only one of a number of influential mathematicians who participated in this, he was one of the most active and successful researchers.

(9) Frobenius's most famous mathematical achievement is to be found in the theory of group characters (for the history of representation theory see [32], [33], and [37]). In 1896 he initiated representation theory of finite groups (over the field **C** of complex numbers) and established what we now conceive as its most basic vocabulary and results [17], [18], [19]. His starting point had been the problem of determining the factorization of a polynomial with integral coefficients – the so-called group determinant – which he learned from Dedekind. Characters first emerged as coefficients of certain terms of the irreducible factors of the group determinant. Already in 1897 Frobenius presented, once again stimulated by Dedekind, the now familiar terminology of matrix representations [20]. The regular representation was proved to be completely reducible and characters turned out to be traces of irreducible matrix representations. The complete reducibility theorem for any representation was given in [23].

Frobenius's further investigations in representation theory seem to have been motivated by the study of the symmetric and the alternating group. To this end he introduced the methods of what is today called induced characters [21] and tensor products of representations [22]. At the turn of the century he managed to determine the characters of the symmetric and alternating groups [24], [25].

Generally, Frobenius worked within the context of existing mathematical disciplines and research traditions. For instance, he neither proposed an entirely different conceptual orientation of a mathematical field, nor did he cast a *completely* new light on problems. This may be one of the reason why Frobenius, in today's estimation, is not considered to be one of the outstanding mathematicians of his age. However, in nearly every one of his approximately one hundred articles, he extended and filled out the realm of mathematical knowledge, and many of his results are considered to be very important achievements of their time. He fairly often provided simplifications and new systematizations of a subject and thus published fundamental, comprehensive expositions which were sometimes used for decades thereafter. At the same time Frobenius's memoirs are rigorous, lucid, and elegant. In this respect he reminds one of his mentor Weierstrass (for Weierstrass's influence on Frobenius see [38]).

The most striking aspect of his mathematical practice is his extraordinary skill at calculations. In fact, Frobenius tried to solve mathematical problems to a large extent by means of a calculative, algebraic approach (but cf. [15]). Even his analytical work was guided by algebraic and linear algebraic methods. For Frobenius, conceptual argumentation played a somewhat secondary role. Although he mostly argued in a comparatively abstract setting, abstraction was not an end in itself (see, e.g., [10], p. 269). Its advantage to him seemed to lie primarily in the fact that it can lead to much greater clearness and precision.

Issai Schur was born in Mogilev, then Russia (the following biographical details rely on [2], pp. 162–237 passim, p. 330, and [3]). He studied exclusively in Berlin from 1894 on where he received his doctorate under Frobenius's supervision in 1901 and where he also habilitated in 1903. From the outset, Schur earned Frobenius's highest respect. With the exception of three years in Bonn (1913–1916) as associate professor, he spent his entire academic life at the University of Berlin. Despite the fact that since 1904 his name had appeared on no fewer than nine shortlists of the Philosophical Faculty for professorships – usually at the instigation of Frobenius – and despite the fact that Frobenius had indicated that Schur was to be his successor, he was not appointed to a chair until 1921 when he became Schottky's successor. At about 1910 Frobenius wrote that "as a Jew and a Russian he unfortunately has few prospects of being accepted by a Prussian university" ([13], p. 18). In 1935 Schur was forced to retire. Four years later he emigrated to Palestine, where he died of heart failure in 1941.

Schur's main lectures dealt with determinant theory, algebraic equations, invariant theory, and algebraic and analytic number theory (for Schur as a teacher see [12], pp. 119, 120, [42], pp. 104, 105, and [53], p. 126). They were thoroughly prepared and exceptionally clear. Unlike Frobenius, Schur placed great importance on being understood and on arousing the interest of his students. Although his lectures were quite demanding (see, e.g., [5], vol. 1, p. xvi), he seems to have succeeded in fulfilling both of these aims in a most outstanding way. Schur was personally interested in students, friendly, and helpful. In fifteen years, he supervised to completion twenty-one or twenty-two dissertations (sources vary: see [2], pp. 359–362, and [95], vol. 3, pp. 479, 480); on average this is about three times more than Frobenius (ten dissertations in twenty-five years).

In the 1920s, Berlin mathematics regained much of its old renown (see [2], pp. 194, 195). Schur's teaching activities contributed considerably to this revival. The development was in part also due to institutional innovations such as the foundation of an Institute for Applied Mathematics. Schur, who like Frobenius was a wholehearted exponent of the Berlin tradition of pure mathematics, does not seem to have participated in those innovations. However, it seems that he did not oppose them – something that Frobenius would probably have done in a number of cases (see, e.g., [2], p. 189).

(1) Schur's most important mathematical contributions are to be found in the field of representation theory (for more details see [3], [41], pp. 153, 154, [42], and [45], pp. 549–552). In 1905 he published a new approach to Frobenius's theory [62], which has remained the standard one till today. Unlike earlier expositions, Schur's reformulation was not based on determinant arguments or on other extraneous theories but only on elementary results on matrices. Its starting point was the result which is now called Schur's lemma.

Schur also generalized Frobenius's theory in several directions. These generalizations exemplify Schur's scholarship at its best. Although no significant preliminary work was carried out by other mathematicians, Schur, in nearly all cases, solved the problems involved completely and with the greatest clarity. Today, his generalizations are at the core of representation theory, and many of his methods have become fundamental to the theory.

Schur's dissertation [58] was on the polynomial representations of the (infinite!) complex general linear group $\mathrm{GL}(n, \mathbf{C})$. (Schur's main results were anticipated by nearly a decade in a paper of Deruyts; see [31].) In the 1920s he rederived the results of his dissertation in a new and briefer way [87], and he also determined the continuous representations of $\mathrm{GL}(n, \mathbf{C})$ as well as of some subgroups of $\mathrm{GL}(n, \mathbf{C})$ [89].

In 1906 Schur treated the arithmetic properties of representations, i.e., he considered representations over "domains of rationality" (e.g., algebraic number fields) which are not necessarily the complex numbers [64]. To this end, he initiated the theory of indices for characters (for other investigations of arithmetic properties see [27], [66], [82]). At about the same time Schur tackled the very difficult problem of the complex projective representations of finite groups [61], [65]. To solve this problem he developed the concept of the multiplier of a group. Later Schur determined the irreducible projective representations of the symmetric and alternating groups [71] and answered the question as to which of these representations are real [88].

Finally, in 1924 Schur determined all the irreducible continuous representations of the real orthogonal and the rotation groups using Hurwitz's method of group integration [84], [85], [86]. The investigation was stimulated by the "counting problem" of invariant theory (see [39], in particular pp. 1652–1655).

(2) In the area of linear substitutions he published a series of articles in which representation theory, if it appeared at all, was only used to lend support (on the understanding of linear homogeneous substitutions at about this time, see [43], pp. 222–232). Schur was interested, for example, in the reducibility and the periodicity of groups of linear substitutions [4], [26], [67], [70], [73], in characteristic roots [68], Hermitian matrices and forms [81], and commutative groups of matrices [60], [63].

(3) In papers on algebraic equations he studied the location of roots (e.g., [80]), problems concerning irreducibility (e.g., [91]), and the Galois group of algebraic equations (e.g., [83]). Schur, for example, was the first mathematician to be able to give equations whose Galois groups are the alternating groups [92], [93].

(4) In one of his few articles on pure group theory he presented a new proof of a theorem first deduced by Frobenius and introduced a homomorphism which is now called the transfer [59].

(5) In the area of analysis Schur published approximately fifteen articles mainly on integral equations (e.g., [68], [69], [75]) and power series (e.g., [74], [78], [79]). What is striking is that even in analysis he – like Frobenius – mainly used and developed what may be called algebraic methods, for the most part from the theories of matrices and linear substitutions.

(6) Finally, Schur also worked on number theory. There are articles on the distribution of prime numbers (e.g., [72]), on algebraic number theory (e.g., [90], [94]) and, most noteworthy, on additive number theory (e.g., [76], [77]). Within the framework of additive number theory, Schur introduced some basic ideas of what quite recently has emerged as a subdiscipline of combinatorial analysis called Ramsey theory (see [98]).

From the outset, Schur was profoundly influenced by Frobenius. The differences between Frobenius and Schur as to their preferred mathematical topics, mathematical languages, methodological techniques and even style of writing are, in fact, amazingly minor. Schur seems to have been somewhat less versatile and his skill at calculations was less marked. On the other hand, in his approximately eighty publications he placed even more emphasis than Frobenius on lucid, attractive presentation and elegant, most appropriate proofs (see [42], p. 102). Schur did not put as much effort as Frobenius into new systematizations of a subject; his strong point was the development of new, well-thought-out ideas and concepts emerging from individual problems. Schur's papers are "always profound and masterly in their restraint" ([13], p. 18) as Frobenius put it.

Schur's understanding of algebra was traditional and strongly determined by the language of polynomials. Throughout his life, he was consistently more interested in algebraic equations, invariant theory, determinants, and, above all, in linear homogeneous substitutions than in abstract algebra and even pure group theory (one of Frobenius's preferred topics). Schur considered the extent of abstraction in modern algebra, especially since the 1920s, to be exaggerated (see [42], pp. 102, 103). Like Frobenius, he was more interested in concrete problems as they occurred within the context of nineteenth century algebra.

References

[1] Mathematische Abhandlungen, Hermann Amandus Schwarz zu seinem fünfzigjährigen Doktorjubiläum am 6. August 1914 gewidmet von Freunden und Schülern. Mit dem Bildnis von H. A. Schwarz und 53 Figuren im Text, Springer-Verlag, Berlin, 1914.

[2] *Biermann, K.-R.*, Die Mathematik und ihre Dozenten an der Berliner Universität 1810–1933, Akademie Verlag, Berlin, 1988.

[3] *Brauer, A.*, Gedenkrede auf Issai Schur, in: [95], vol. 1, V–XIV.

[4] *Brauer, R., Schur, I.*, Zum Irreduzibilitätsbegriff in der Theorie der Gruppen linearer homogener Substitutionen, Sitzungsberichte der Preussischen Akademie der Wissenschaften, Physikalisch-Mathematische Klasse (1930), 209–226.

[5] *Brauer, R.*, Collected Papers, 3 vols., Fong, P., Wong, W.J. (eds.), The MIT Press, Cambridge, Massachusetts, and London, 1980.

[6] *Brewer, J.W., Smith, M.K. (eds.)*, Emmy Noether, A Contribution to Her Life and Work, Dekker, New York, Basel, 1981.

[7] *vom Brocke, B.*, Wissenschaftsgeschichte und Wissenschaftspolitik im Industriezeitalter, Das "System Althoff" in historischer Perspektive, Geschichte von Bildung und Wissenschaft, Reihe B, Sammelwerke 5, Lax, Hildesheim, 1991.

[8] *Casadio, G., Zappa, G.*, Storia del teorema di Sylow e delle sue dimostrazioni, Bollettino di Storia delle Scienze Matematiche **10** (1990), 29–75.

[9] *Dedekind, R.*, Gesammelte mathematische Werke, 2 vols., Fricke, R., Noether, E., Ore, Ö. (eds.), Vieweg, Braunschweig, 1931.

[10] *Dugac, P.*, Richard Dedekind et les fondements des mathématiques, Préface de Jean Dieudonné, L'Histoire des Sciences, Textes et Études, Collection des Travaux de l'Académie internationale d'Histoire des Sciences 24, Librairie Philosophique J. Vrin, Paris, 1976.

[11] *Ferreirós, J.*, Labyrinth of reason, A history of set theory and its role in modern mathematics, Birkhäuser, Basel, Boston, Berlin, 1998.

[12] *Fraenkel, A.A.*, Lebenskreise, Aus den Erinnerungen eines jüdischen Mathematikers, Deutsche Verlags-Anstalt, Stuttgart, 1967.

[13] *Frei, G., Stammbach, U.*, Die Mathematiker an den Züricher Hochschulen, Birkhäuser, Basel, Boston, Berlin, 1994.

[14] *Frobenius, G., Stickelberger, L.*, Ueber Gruppen von vertauschbaren Elementen, J. Reine Angew. Math. **86** (1879), 217–262.

[15] *Frobenius, G.*, Antrittsrede, Sitzungsberichte der Königlich Preussischen Akademie der Wissenschaften zu Berlin (1893), 626–628.

[16] *Frobenius, G.*, Über Beziehungen zwischen den Primidealen eines algebraischen Körpers und den Substitutionen seiner Gruppe, Sitzungsberichte der Königlich Preussischen Akademie der Wissenschaften zu Berlin (1896), 689–703.

[17] *Frobenius, G.*, Über Gruppencharaktere, Sitzungsberichte der Königlich Preussischen Akademie der Wissenschaften zu Berlin (1896), 985–1021.

[18] *Frobenius, G.*, Über die Primfactoren der Gruppendeterminante, Sitzungsberichte der Königlich Preussischen Akademie der Wissenschaften zu Berlin (1896), 1343–1382.

[19] *Frobenius, G.*, Über vertauschbare Matrizen, Sitzungsberichte der Königlich Preussischen Akademie der Wissenschaften zu Berlin (1896), 601–614.

[20] *Frobenius, G.*, Über die Darstellung der endlichen Gruppen durch lineare Substitutionen, Sitzungsberichte der Königlich Preussischen Akademie der Wissenschaften zu Berlin (1897), 994–1015.

[21] *Frobenius, G.*, Über Relationen zwischen den Charakteren einer Gruppe und denen ihrer Untergruppen, Sitzungsberichte der Königlich Preussischen Akademie der Wissenschaften zu Berlin (1898), 501–515.

[22] *Frobenius, G.*, Über die Composition der Charaktere einer Gruppe, Sitzungsberichte der Königlich Preussischen Akademie der Wissenschaften zu Berlin (1899), 330–339.

[23] *Frobenius, G.*, Über die Darstellung der endlichen Gruppen durch lineare Substitutionen II, Sitzungsberichte der Königlich Preussischen Akademie der Wissenschaften zu Berlin (1899), 482–500.

[24] *Frobenius, G.*, Über die Charaktere der symmetrischen Gruppe, Sitzungsberichte der Königlich Preussischen Akademie der Wissenschaften zu Berlin (1900), 516–534.

[25] *Frobenius, G.*, Über die Charaktere der alternierenden Gruppe, Sitzungsberichte der Königlich Preussischen Akademie der Wissenschaften zu Berlin (1901), 303–315.

[26] *Frobenius, G., Schur, I.*, Über die Äquivalenz der Gruppen linearer Substitutionen, Sitzungsberichte der Königlich Preussischen Akademie der Wissenschaften zu Berlin (1906), 209–217.

[27] *Frobenius, G., Schur, I.*, Über die reellen Darstellungen der endlichen Gruppen, Sitzungsberichte der Königlich Preussischen Akademie der Wissenschaften zu Berlin (1906), 186–208.

[28] *Frobenius, G.*, Gesammelte Abhandlungen, 3 vols., Serre, J-P. (ed.), Springer-Verlag, Berlin, Heidelberg, New York, 1968.

[29] *Grattan-Guinness, I.*, Companion Encyclopedia of the History and Philosophy of the Mathematical Sciences, 2 vols., Routledge, London, New York, 1994.

[30] *Gray, J.*, Linear Differential Equations and Group Theory from Riemann to Poincaré, Birkhäuser, Boston, Basel, Stuttgart, 1986.

[31] *Green, J.A.*, Classical Invariants and the General Linear Group, in: [46], 247–272.

[32] *Hawkins, T.*, The Origins of the Theory of Group Characters, Arch. Hist. Exact Sci. **7** (1970/71), 142–170.

[33] *Hawkins, T.*, New Light on Frobenius' Creation of the Theory of Group Characters, Arch. Hist. Exact Sci. **12** (1974), 217–243.

[34] *Hawkins, T.*, The Theory of Matrices in the 19th Century, in: [50], vol. 2, 561–570.

[35] *Hawkins, T.*, Another Look at Cayley and the Theory of Matrices, Archives internationales d'histoire des sciences **27** (1977), 82–112.

[36] *Hawkins, T.*, Weierstrass and the Theory of Matrices, Arch. Hist. Exact Sci. **17** (1977), 119–163.

[37] *Hawkins, T.*, The creation of the theory of group characters, Rice University Studies **64** (1978), Nos. 2 & 3, 57–71.

[38] *Hawkins, T.*, The Berlin school of mathematics, in: [44], 233–245.

[39] *Hawkins, T.*, Cayley's Counting Problem and the Representation of Lie Algebras, in: [51], vol. 2, 1642–1656.

[40] Ein Jahrhundert Mathematik, 1890–1990, Festschrift zum Jubiläum der DMV, Dokumente zur Geschichte der Mathematik 6, Vieweg, Braunschweig, Wiesbaden, 1990.

[41] *Lam, T.Y.*, Representation Theory, in: [6], 145–156.

[42] *Ledermann, W.*, Issai Schur and his school in Berlin, Bull. London Math. Soc. **15** (1983), 97–106.

[43] *Loewy, A.*, Algebraische Gruppentheorie, in: [49], 168–249.

[44] *Mehrtens, H., Bos, H., Schneider, I.*, Social History of Nineteenth Century Mathematics, Birkhäuser, Boston, Basel, Stuttgart, 1981.

[45] *Michler, G.O.*, Vom Hilbertschen Basissatz bis zur Klassifikation der endlichen einfachen Gruppen, in: [40], 537–586.

[46] *Michler, G.O., Ringel, C.M.*, Representation Theory of Finite Groups and Finite-Dimensional Algebras, Progress Math. 95, Birkhäuser, Basel, Boston, Berlin, 1991.

[47] *Miller, G.A.*, Historical sketch of the development of the theory of groups of finite order, Bibliotheca Mathematica, Zeitschrift für Geschichte der mathematischen Wissenschaften 3. Folge **10** (1909/10), 317–329.

[48] *Nicholson, J.*, The Development and Understanding of the Concept of Quotient Group, Hist. Math. **20** (1993), 68–88.

[49] *Pascal, E.*, Repertorium der höheren Analysis, 2nd edition, 1st part, Algebra, Differential- und Integralrechnung, Teubner, Leipzig und Berlin, 1910.

[50] Proceedings of the International Congress of Mathematicians, 2 vols., Vancouver, 1975.

[51] Proceedings of the International Congress of Mathematicians, August 3–11, 1986, 2 vols., Berkeley, 1987.

[52] *Pyenson, L.*, Neohumanism and the Persistence of Pure Mathematics in Wilhelmian Germany, American Philosophical Society, Philadelphia, 1983.

[53] *Rohrbach, H.*, Richard Brauer zum Gedächtnis, Jber. dt. Math.-Verein. **83** (1981), 125–134.

[54] *Rowe, D.E.*, Klein, Hilbert, and the Göttingen Mathematical Tradition, Osiris, Second Series **5** (1989), 186–213.

[55] *Schlesinger, L.*, Bericht über die Entwickelung der Theorie der linearen Differentialgleichungen seit 1865, Jber. dt. Math.-Verein. **18** (1909), 133–266.

[56] *Schneider, H.*, The Concepts of Irreducibility and Full Indecomposability of a Matrix in the Works of Frobenius, König and Markov, Linear Algebra and Its Applications **18** (1977), 139–162.

[57] *Schubring, G.*, Germany to 1933, in: [29], vol. 2, 1442–1456.

[58] *Schur, I.*, Ueber eine Klasse von Matrizen, die sich einer gegebenen Matrix zuordnen lassen, Diss., Berlin, 1901.

[59] *Schur, I.*, Neuer Beweis eines Satzes über endliche Gruppen, Sitzungsberichte der Königlich Preussischen Akademie der Wissenschaften zu Berlin (1902), 1013–1019.

[60] *Schur, I.*, Über einen Satz aus der Theorie der vertauschbaren Matrizen, Sitzungsberichte der Königlich Preussischen Akademie der Wissenschaften zu Berlin (1902), 120–125.

[61] *Schur, I.*, Über die Darstellung der endlichen Gruppen durch gebrochene lineare Substitutionen, J. Reine Angew. Math. **127** (1904), 20–50.

[62] *Schur, I.*, Neue Begründung der Theorie der Gruppencharaktere, Sitzungsberichte der Königlich Preussischen Akademie der Wissenschaften zu Berlin (1905), 406–432.

[63] *Schur, I.*, Zur Theorie der vertauschbaren Matrizen, J. Reine Angew. Math. **130** (1905), 66–76.

[64] *Schur, I.*, Arithmetische Untersuchungen über endliche Gruppen linearer Substitutionen, Sitzungsberichte der Königlich Preussischen Akademie der Wissenschaften zu Berlin (1906), 164–184.

[65] *Schur, I.*, Untersuchungen über die Darstellung der endlichen Gruppen durch gebrochene lineare Substitutionen, J. Reine Angew. Math. **132** (1907), 85–137.

[66] *Schur, I.*, Über die Darstellung der symmetrischen Gruppe durch lineare homogene Substitutionen, Sitzungsberichte der Königlich Preussischen Akademie der Wissenschaften zu Berlin (1908), 664–678.

[67] *Schur, I.*, Beiträge zur Theorie der Gruppen linearer homogener Substitutionen, Trans. Am. Math. Soc. **10** (1909), 159–175.

[68] *Schur, I.*, Über die charakteristischen Wurzeln einer linearen Substitution mit einer Anwendung auf die Theorie der Integralgleichungen, Math. Ann. **66** (1909), 488–510.

[69] *Schur, I.*, Zur Theorie der linearen homogenen Integralgleichungen, Math. Ann. **67** (1909), 306–339.

[70] *Schur, I.*, Über Gruppen periodischer linearer Substitutionen, Sitzungsberichte der Königlich Preussischen Akademie der Wissenschaften zu Berlin (1911), 619–627.

[71] *Schur, I.*, Über die Darstellung der symmetrischen und der alternierenden Gruppe durch gebrochene lineare Substitutionen, J. Reine Angew. Math. **139** (1911), 155–250.

[72] *Schur, I.*, Über die Existenz unendlich vieler Primzahlen in einigen speziellen arithmetischen Progressionen, Sitzungsberichte der Berliner Mathematischen Gesellschaft **11** (1912), 40–50.

[73] *Schur, I.*, Über Gruppen linearer Substitutionen mit Koeffizienten aus einem algebraischen Zahlkörper, Math. Ann. **71** (1912), 355–367.

[74] *Schur, I.*, Über einen Satz von C. Carathéodory, Sitzungsberichte der Königlich Preussischen Akademie der Wissenschaften zu Berlin (1912), 4–15.

[75] *Schur, I.*, Über die Entwicklung einer gegebenen Funktion nach den Eigenfunktionen eines positiv definiten Kerns, in: [1], 392–409.

[76] *Schur, I.*, Ein Beitrag zur additiven Zahlentheorie und zur Theorie der Kettenbrüche, Sitzungsberichte der Königlich Preussischen Akademie der Wissenschaften zu Berlin (1917), 302–321.

[77] *Schur, I.*, Über die Kongruenz $x^m + y^m \equiv z^m \pmod{p}$, Jber. dt. Math.-Verein. **25** (1916), 114–117.

[78] *Schur, I.*, Über Potenzreihen, die im Innern des Einheitskreises beschränkt sind, J. Reine Angew. Math. **147** (1917), 205–232.

[79] *Schur, I.*, Über Potenzreihen, die im Innern des Einheitskreises beschränkt sind (Fortsetzung), J. Reine Angew. Math. **148** (1918), 122–145.

[80] *Schur, I.*, Über die Verteilung der Wurzeln bei gewissen algebraischen Gleichungen mit ganzzahligen Koeffizienten, Math. Z. **1** (1918), 377–402.

[81] *Schur, I.*, Über endliche Gruppen und Hermitesche Formen, Math. Z. **1** (1918), 184–207.

[82] *Schur, I.*, Einige Bemerkungen zu der vorstehenden Arbeit des Herrn A. Speiser, Math. Z. **5** (1919), 7–10.

[83] *Schur, I.*, Beispiele für Gleichungen ohne Affekt, Jber. dt. Math.-Verein. **29** (1920), 145–150.

[84] *Schur, I.*, Neue Anwendungen der Integralrechnung auf Probleme der Invariantentheorie, Sitzungsberichte der Preussischen Akademie der Wissenschaften, Physikalisch-Mathematische Klasse (1924), 189–208.

[85] *Schur, I.*, Neue Anwendungen der Integralrechnung auf Probleme der Invariantentheorie II, Über die Darstellung der Drehungsgruppe durch lineare homogene Substitutionen, Sitzungsberichte der Preussischen Akademie der Wissenschaften, Physikalisch-Mathematische Klasse (1924), 297–321.

[86] *Schur, I.*, Neue Anwendungen der Integralrechnung auf Probleme der Invariantentheorie III, Vereinfachung des Integralkalküls, Realitätsfragen, Sitzungsberichte der Preussischen Akademie der Wissenschaften, Physikalisch-Mathematische Klasse (1924), 346–355.

[87] *Schur, I.*, Über die rationalen Darstellungen der allgemeinen linearen Gruppe, Sitzungsberichte der Preussischen Akademie der Wissenschaften, Physikalisch-Mathematische Klasse (1927), 58–75.

[88] *Schur, I.*, Über die reellen Kollineationsgruppen, die der symmetrischen oder der alternierenden Gruppe isomorph sind, J. Reine Angew. Math. **158** (1927), 63–79.

[89] *Schur, I.*, Über die stetigen Darstellungen der allgemeinen linearen Gruppe, Sitzungsberichte der Preussischen Akademie der Wissenschaften, Physikalisch-Mathematische Klasse (1928), 100–124.

[90] *Schur, I.*, Elementarer Beweis eines Satzes von L. Stickelberger, Math. Z. **29** (1929), 464–465.

[91] *Schur, I.*, Zur Irreduzibilität der Kreisteilungsgleichung, Math. Z. **29** (1929), 463.

[92] *Schur, I.*, Gleichungen ohne Affekt, Sitzungsberichte der Preussischen Akademie der Wissenschaften, Physikalisch-Mathematische Klasse (1930), 443–449.

[93] *Schur, I.*, Affektlose Gleichungen in der Theorie der Laguerreschen und Hermiteschen Polynome, J. Reine Angew. Math. **165** (1931), 52–58.

[94] *Schur, I.*, Einige Bemerkungen über die Diskriminante eines algebraischen Zahlkörpers, J. Reine Angew. Math. **167** (1932), 264–269.

[95] *Schur, I.*, Gesammelte Abhandlungen, 3 vols., Brauer, A., Rohrbach, H. (eds.), Springer-Verlag, Berlin, Heidelberg, New York, 1973.

[96] *Siegel, C.L.*, Erinnerungen an Frobenius, in: [28], vol. 1, IV–VI.

[97] *Silvestri, R.*, Simple Groups of Finite Order in the Nineteenth Century, Arch. Hist. Exact Sci. **20** (1979), 313–356.

[98] *Soifer, A.*, Issai Schur, Ramsey theory before Ramsey, Combinatorics **5** (1995), 6–23.

[99] *Tobies, R.*, Wissenschaftliche Schwerpunktbildung, der Ausbau Göttingens zum Zentrum der Mathematik und Naturwissenschaften, in: [7], 87–108.

[100] *Waterhouse, W.C.*, The Early Proofs of Sylow's Theorem, Arch. Hist. Exact Sci. **21** (1979/80), 279–290.

[101] *Wussing, H.*, The Genesis of the Abstract Group Concept, A Contribution to the History of the Origin of Abstract Group Theory, The MIT Press, Cambridge, London, 1984.

Erhard Schmidt, John von Neumann

Hellmut Baumgärtel

Biographical notes

Erhard Schmidt was born on January 13, 1876, in Dorpat. He achieved the position of *Ordinarius* (full professor) at the University of Berlin (then the *Friedrich-Wilhelms-Universität*) in 1917, succeeding H.A. Schwarz. Schmidt remained *Ordinarius* until 1950. As co-director of the mathematics department (a position he held from 1917 to 1952) he was involved in the founding of the new Institute for Applied Mathematics in 1918 and thus played a major role in establishing this discipline at the university. His commitment was not confined to his own department, as is attested to by his serving as dean (1921/22) and vice-chancellor (*Rektor*, 1929/30) of the university. It was at an inaugural speech for this last office that he chose as his topic *Über Gewißheit in der Mathematik* (On certainty in mathematics) ([A1]; [C5], pp. 221, 222). From 1946 until 1958, he acted as director at the newly founded research institute for mathematics of the German Academy of Sciences and in 1948 he was a co-founder and first editor of the journal *Mathematische Nachrichten*.

With the rise of the Nazi regime in 1933, many of his colleagues and friends were forced into exile – I. Schur, R. v. Mises, B. Neumann, A. and R. Brauer, among others, had to leave the country. Freudenthal left Germany already in 1931. He got a position at the University of Amsterdam. However, when the Netherlands were occupied by the Germans, he had to leave the university. For a time he was put into prison but he escaped from deportation.

Erhard Schmidt stayed in Berlin. His attitude towards national socialism commanded great respect, even among his exiled colleagues. This is reflected in H. Freudenthal's address on the occasion of Schmidt's 75th birthday:

"*Es klingt wohl etwas merkwürdig, wenn man jemandem ausdrücklich bescheinigt, daß er ein ehrlicher Mann ist, und doch wissen wir alle, daß Jahre hinter uns liegen, in denen es einem schwer gemacht wurde, ein ehrlicher Mensch zu sein. Ach ja, es ist ja so leicht, die Ehrlichkeit, die die Mathematik erfordert, in der Mathematik zu üben, denn tut man es nicht, so rächt es sich schnell und bitter. Es ist so viel schwerer, dieser an Zahlen und Figuren erprobten Charaktereigenschaft auch treu zu bleiben unter den Menschen und gegenüber den Freunden. Daß wir draußen, denen Deutschland jahrelang verschlossen und Feind war, das wissen und daß wir nie [an Erhard Schmidt] gezweifelt haben, beweist die große Zahl der Beiträge, die von dort die Herausgeber der Festschrift [zum 75. Geburtstag von E. Schmidt] erreicht haben.*" ("It may sound a bit odd to explicitly vouch for somebody that he is an honest and upright man. But we all know that during the years behind us it was difficult to be an honest und upright man. Alas, it is so easy to practice the kind of honesty needed in mathematics inside mathematics, for if one does not, one soon has to pay dearly. It is so much more difficult to stick

Erhard Schmidt

to this trait of character, exercised on numbers and figures, also with respect to men and friends. The fact that we, who have been abroad and to whom Germany was sealed and hostile for years, know this, but have never had any doubts [about Erhard Schmidt], is testified by the great number of contributions to the *Festschrift* [on the occasion of E. Schmidt's 75th birthday] that the editors have received from there.") ([C1] quoted from [C7], p. 238)

In his response, Erhard Schmidt coined the phrase "*von der Würde nach außen, die man nicht verlieren, und von der Liebe nach innen, in der man nicht nachlassen dürfe*" ("of the dignity one must not lose outwardly, and the love inside in which one must not slacken"). At a time of great shame for those German mathematicians who forgot the dignity of their fellow men and their own, Erhard Schmidt felt bound to oppose these misdemeanors in his own way ([C5], p. 218). In his own words, spoken on the same occasion:

"*Ich liebte eben meine Schüler. Genau dasselbe gilt auch für die Universität als Ganzes. Ich liebe die Berliner Universität, ob sie sich nun in glücklicherer Lage befindet oder nicht – das ändert daran nichts. Ich liebe sie, seit ich in Berlin bin, und werde ihr Treue halten.*" ("I simply loved my students. And exactly the same is true of the university as a whole. I love the University of Berlin, whether it happens to be in happy conditions or not – this does not change anything. I have loved it from the time I have been in Berlin and I will remain faithful to it.") ([C1] quoted from [C7], p. 238)

On the other hand it should be noted that there are also critical statements on Schmidt's attitude in that time. We mention his approval of the following statement of the *Berliner Akademie der Wissenschaften*, when Einstein left the academy (on March 18, 1933):

"*Die Preußische Akademie der Wissenschaften empfindet das agitatorische Auftreten Einsteins im Auslande umso schwerer, als sie und ihre Mitglieder seit alten Zeiten sich aufs engste mit dem preußischen Staat verbunden fühlt und ... den nationalen Gedanken stets betont und bewahrt hat. Sie hat aus diesem Grunde keinen Anlaß, den Austritt Einsteins zu bedauern.*" ("The Prussian Academy of Sciences is struck all the more by Einstein's rabble-rousing declarations abroad as it and its members have always felt a close loyalty to the Prussian state, and ... emphasized and conserved the national idea. It therefore sees no reason to regret Einstein's resignation.") (quotation from [C11])

In 1936, Erhard Schmidt was chairman and speaker of the German delegation to the International Congress of Mathematicians at Oslo.

Erhard Schmidt, a bright star in the mathematical sky of Berlin, passed away December 6, 1959.

Among those contributors to the *Festschrift* ([C2]), mentioned by Freudenthal, above, who had been in exile was John von Neumann.

John (Johann) von Neumann was born on December 28, 1903, in Budapest. His time at the University of Berlin, spanning the years 1927 to 1933 was brief, but extraordinarily productive. "*Von Anfang an erfreute er sich des förderlichen Interesses Erhardt Schmidts, der in ihm einen Partner für tiefe fachliche Gespräche fand.*" ("From the beginning, he enjoyed the patronizing interest of Erhard Schmidt, who found in him a partner for deep specialized conversation.") (quotation from [C7], p. 223)

V. Neumann had stayed in Berlin before in 1921, at a time when he was a student of chemical engineering. 1927 was the year of his *Habilitation*, his qualification as a lecturer in Berlin. For his *Habilitationsschrift*, the thesis paper associated with the qualification process, Erhard Schmidt was the principal examiner. As early as the summer term of 1928, Neumann started lecturing in Berlin. Starting in the winter term 1928/29, he got a position as *Privatdozent* to lecture on the logical basis of mathematics (*Vorlesungen über die logischen Grundlagen der Mathematik*). The summer term 1929 he spent in Hamburg, on leave of absence from Berlin. During the summer term 1929 as well as the winter tems in 1931/32 and 1932/33, he stayed in the USA. All in all, he lectured in Berlin for five terms. Due to the rise of the Nazi regime, he was not able to give the lectures he had planned for the summer term 1933 ([C7], p. 209).

His last encounter with Erhard Schmidt was in February 1933, in Berlin ([C8], p. 149).

Even this brief description demonstrates the "cometary" nature of his work in Berlin. Brief though his stay may have been, it left lasting scientific traces. Many of the most important results we associate with von Neumann's name were found during this time, as can be seen from his publications (see [B1]–[B8]).

On the work of Erhard Schmidt

Erhard Schmidt's contributions to mathematics concern the areas of integral equations, Hilbert spaces, partial differential equations and geometry (isoperimetric inequalities). Mainly in the twenties, he was also interested in topology. It was then that he developed a today classic proof of Jordan's theorem of curves ([C5], p. 215). In 1929, he was the main examiner for Heinz Hopf's dissertation, *Über Zusammenhänge zwischen Topologie und Metrik von Mannigfaltigkeiten*. Hopf, who had been fascinated by Erhard Schmidt's 1917 lectures on set theory, had followed his admired teacher to the University of Berlin in 1920 ([C7], pp. 335–337; [C6], p. 182 quoted from [C5], pp. 215, 216).

Integral equations: Here, Erhard Schmidt continued the investigations by Hilbert and Fredholm. In his work [A2] (which reiterates the main points of his Göttingen dissertation), the theory of linear integral equations is developed from very simple premises, rendering the results easily generalizable. In his paper [A4], he presents a very elementary and clear derivation of Fredholm's principal theorems on the solution of linear integral equations. It is in these works that we find the orthogonalization method known today as the Gram-Schmidt process and the series expansion (of a compact operator) that, nowadays, is referred to as the Schmidt series. In [A4], solution methods for non-linear integral equations are treated.

Hilbert space: Influenced by Hilbert's idea to introduce the notions and concepts of geometry into the theory of function spaces, Erhard Schmidt explicitly developed in [A5] the concept of a Hilbert space and the geometry of such a space by means of the space l^2. The Gram-Schmidt orthogonalization, the theorem of projections and the treatment of those infinite matrices that correspond to the Hilbert-Schmidt operators of today's terminology can be found here. As this research shows, Schmidt was one of the first mathematicians to show that the concepts of Euclidean geometry can be generalized towards establishing abstract theories that have important applications (see [C3], pp. 189, 190).

On the work of John von Neumann

Logic, set theory, measure theory, the theory of Hilbert spaces (especially the spectral theory of operators) and mathematical physics (especially quantum mechanics) all formed part of John von Neumann's work during his time in Berlin. Logic and set theory are the topic of his professorial thesis (*Habilitationsschrift*) and the lectures he gave. In this review, we will concentrate on his contributions to mathematical physics and to the theory of Hilbert spaces. In the latter area, the connection between his work and that of Erhard Schmidt is especially clear.

It is in his paper [B1] that von Neumann formulates for the first time the notion of the abstract Hilbert space: "*Somit entsteht für uns die Aufgabe, den abstrakten Hilbertraum auf Grund seiner 'inneren' – d.h. ohne die Bezugnahme auf die Interpretation seiner Elemente als Folgen oder Funktionen formulierbaren – Eigenschaften zu beschreiben.*" ("Thus arises the task to describe the abstract

John von Neumann

Hilbert space according to its intrinsic properties, i.e., without referring to an interpretation of its elements as series or functions.") In his works [B2], [B4], [B6], the theory of unbounded operators and the theory of extensions of symmetric operators is developed in a way that still remains the standard formulation of the theory today. The paper [B3] contains the double commutant theorem for algebras known today as von Neumann algebras and the spectral theory of unbounded operators, especially the spectral theorem for normal operators. [B5] describes the functional calculus for self-adjoint operators. In [B7], a cornerstone of quantum mechanics can be found: the uniqueness theorem for regular irreducible representations of the algebra of canonical commutation relations for systems with finitely many degrees of freedom. In his famous book [B8], among other things, the mathematical apparatus for quantum mechanics is summarized.

It should be stressed that "... within a few years [von Neumann] realized that the traditional idea of representing an operator by an infinite matrix was totally inadequate..." ([C4], p. 90), an inadequacy that became apparent in the so-called "pathological properties" of unbounded hermitian matrices (see [B4]).

Connections between the lives and works of Erhard Schmidt and John von Neumann

The first encounter between the two mathematicians can be dated to 1921, when von Neumann visited Berlin for the first time. We know that, around 1922–23,

Erhard Schmidt sent the first version of John von Neumann's thesis *Die Axiomatik der Mengenlehre* to A. Fraenkel in Marburg, asking him for his opinion. Fraenkel soon recognized, *ex ungue leonem*, what mathematical lion had written it ([C8], pp. 89, 90). In 1927, Erhard Schmidt was the principal examiner for John von Neumann's professorial thesis *Der axiomatische Aufbau der Mengenlehre*.

The connection between the research of von Neumann and that of Schmidt which, from the point of view of both mathematics and the history of mathematics, is the most interesting, concerns the abstract notion of Hilbert space. As mentioned above, Erhard Schmidt had developed the theory of Hilbert spaces, especially their topology and geometry, using the example of l^2. Therefore, "linear operators", in today's terminology, appeared there only as infinite matrices. John von Neumann, before coming to Berlin in 1927, had participated in the struggle for the relationship between, or, more precisely, the equivalence of Schrödinger's and Heisenberg's approach to quantum mechanics, that had taken place in Göttingen. Since the spaces l^2 and the L^2-function space turned out to be isomorphic, there was, so to say, an objective tendency towards the idea of an abstract Hilbert space, which he published in [B1]. This inevitably led to an abstract notion of "linear operator" and also to "unbounded operators" that are important in quantum mechanics and that lead to the realization that the "matrix representation" was, in fact, inexpedient.

In this context, von Neumann's research was obviously stimulated by Erhard Schmidt, e.g., the expression "hypermaximal" for an operator (the property that, later, came to be called self-adjoint), which von Neumann uses in [B2], was coined by Schmidt (see, e.g., [B8], p. 88 and the associated remark [B8], p. 248; see also [C9], resp. a quotation from [C9] in [C5], p. 215 and p. 210). On the other hand, Erhard Schmidt seems to have preferred the notion of "infinite matrices" even after the axiomatization of the Hilbert space concept. This indicates that the new step was only reluctantly accepted (see [C8], p. 128; see also [C10], p. 287). The new, more abstract approach was a great leap ahead, especially as far as the spectral theory of operators was concerned. It caused a flurry of newly written (or rewritten) papers (see, for example, the remarks by K.O. Friedrichs mentioned in [C8], p. 125), and quickly made older works, even books that had just been published, at least partially obsolete.

The last personal encounter between Erhard Schmidt and John von Neumann took place in February 1933. John von Neumann, coming from America and on his way to Budapest, stopped in Berlin. In a letter written in 1949, he expressed his great admiration to Erhard Schmidt for the latter's work [A6] on the isoperimetric problem (cf. [C5], p. 221).

To the collection of essays on the occasion of Schmidt's 75th birthday, John von Neumann contributed a paper [B9] that is "dedicated to Erhard Schmidt's 75th birthday in adoration" ([B9], [C12]).

Acknowledgements

It is a pleasure to thank Dr. R. Bölling (University of Potsdam) for helpful remarks and for pointing out the references [C7], [C3] and [C4], Prof. H. Begehr and Dr. R. Siegmund-Schultze for several critical remarks, and Prof. A. Pietsch (University of Jena) for pointing out the reference [C10]. The author is indebted to R. Helling and M. Pössel (Albert Einstein Institut für Gravitationsphysik Potsdam) for translating a first version into English.

Bibliography

A. Original papers of Erhard Schmidt

[A1] Über Gewißheit in der Mathematik, Rektoratsrede vom 15.10.1929, Berlin, 1930.

[A2] Zur Theorie der linearen und nichtlinearen Integralgleichungen I, Entwicklung willkürlicher Funktionen nach Systemen vorgeschriebener, Math. Ann. **63** (1907), 433–476.

[A3] Zur Theorie der linearen und nichtlinearen Integralgleichungen II, Auflösung der allgemeinen linearen Integralgleichung, Math. Ann. **64** (1907), 161–174.

[A4] Zur Theorie der linearen und nichtlinearen Integralgleichungen III, Über die Auflösung der nichtlinearen Integralgleichung und die Verzweigung ihrer Lösungen, Math. Ann. **65** (1908), 370–399.

[A5] Über die Auflösung linearer Gleichungen mit unendlich vielen Unbekannten, Rendiconti del Circolo matematico di Palermo **25** (1908), 53–77.

[A6] Die Brunn-Minkowskische Ungleichung und ihr Spiegelbild sowie die isoperimetrische Eigenschaft der Kugel in der euklidischen und nichteuklidischen Geometrie II, Math. Nachr. **2** (1949), 171–244.

B. Original papers of John von Neumann

[B1] Mathematische Begründung der Quantenmechanik, Göttinger Nachrichten **1** (9) (1927), 1–57.

[B2] Allgemeine Eigenwerttheorie Hermitescher Funktionaloperatoren, Math. Ann. **102** (1929), 49–131.

[B3] Zur Algebra der Funktionaloperationen und Theorie der normalen Operatoren, Math. Ann. **102** (1929), 370–427.

[B4] Zur Theorie der unbeschränkten Matrizen, J. Reine Angew. Math. **161** (1929), 208–236.

[B5] Über Funktionen von Funktionaloperatoren, Ann. of Math. **32** (1931), 191–226.

[B6] Über adjungierte Funktionaloperatoren, Ann. of Math. **33** (1932), 294–310.

[B7] Die Eindeutigkeit der Schrödingerschen Operatoren, Math. Ann. **104** (1931), 570–578.

[B8] Mathematische Grundlagen der Quantenmechanik, Springer-Verlag, Berlin, 1932.

[B9] Eine Spektraltheorie für allgemeine Operatoren eines unitären Raumes, Math. Nachr. **4** (1950/51), 258–281.

C. Further references

[C1] Ansprachen anläßlich der Feier des 75. Geburtstages von Erhard Schmidt durch seine Fachgenossen (13.01.1951), Vervielfältigung (nicht im Druck erschienen), Berlin, 1951.

[C2] Festband zum 75. Geburtstag von Erhard Schmidt, Math. Nachr. **4** (1950/51), 1–511.

[C3] *Bernkopf, M.*, Schmidt, Erhard, Dictionary of Scientific Biography **12** (1974), 187–190.

[C4] *Dieudonné, J.*, Von Neumann, Johann (or John), Dictionary of Scientific Biography **14** (1976), 88–92.

[C5] *Rohrbach, H.*, Erhard Schmidt, Ein Lebensbild, Jber. dt. Math.-Verein. **69** (1968), 209–224.

[C6] *Hopf, H.*, Ein Abschnitt aus der Entwicklung der Topologie, Jber. dt. Math.-Verein. **68** (1966), 182–192.

[C7] *Biermann, K.-R.*, Die Mathematik und ihre Dozenten an der Berliner Universität 1810–1933, Akademie Verlag, Berlin, 1988.

[C8] *Macrae, N.*, John von Neumann, Birkhäuser, Basel, Boston, Berlin, 1994.

[C9] *Nevanlinna, R.*, Erhard Schmidt zu seinem 80. Geburtstag, Math. Nachr. **15** (1956), 3–6.

[C10] *Pietsch, A.*, Nachwort, in: Hilbert, D., Schmidt, E. (eds.), Integralgleichungen und Gleichungen mit unendlich vielen Unbekannten, BSB Teubner, 1998, 279–311.

[C11] Einstein 1879–1979, Ausstellung, Jüdische National- und Universitätsbibliothek, Berman Saal, Jerusalem, März 1979.

[C12] *Berberian, S.B., Baumgärtel, H.*, Bemerkung zu einem Satz von J. v. Neumann, Math. Nachr. **32** (1966), 313–316.

Constantin Carathéodory (1873–1950)

Heinrich Begehr

It is not handed down but it looks very likely that Carathéodory's decision to accept the call to replace Frobenius in 1918 at the *Friedrich-Wilhelms-Universität* in Berlin was a surprise to many of his contemporaries. Since 1913 he was full professor in Göttingen as the successor of Felix Klein. At that time Göttingen was the leading German university in mathematics.

Several times before, the University of Berlin occasionally supported by the Prussian Academy of Sciences, had tried to hire mathematicians from Göttingen in vain. Gauss did not accept a call in 1810 – and there was no better luck in 1822 – neither did Hilbert in 1902, 1914, 1917 (see [B]). The only exeption was H.A. Schwarz who left Göttingen in 1892 for the Weierstrass chair. In comparison, Göttingen was more lucky getting Dirichlet in 1856 as successor of Gauss and hiring Landau in 1909 replacing Minkowski.

There were some personal reasons for Carathéodory to come to Berlin. He was born in Berlin. His parents both were descendents from famous Greek families. His father was at that time an attaché of the Turkish Embassy in Berlin. Carathéodory grew up in Brussels and attended the *École Militaire* there. His family background and his excellent education is extensively described in [1], [2]. During his first employment as an engineer in 1898 in Egypt, where he was involved in the building of the dams for the Nile River in Asyût and Aswân, he started to read Jordan's *cours d'analyse* and other books in mathematics.

Surprisingly for friends and relatives, he decided to quit his job and to start to study mathematics. Thinking to go to Paris or to some German university, his final choice was Berlin. He started in spring 1900. The German university system which leaves the choice of courses to the students suited Carathéodory's capabilities very well. He immediately came into close contact with H.A. Schwarz who invited him to his colloquium. In this stimulating colloquium where often the own results of Schwarz were reported, Carathéodory met all the other gifted mathematics students, among them, Paul Koebe, Adolf Kneser, Fritz Hartogs, Leopold Fejér and Erhard Schmidt. Carathéodory and Schmidt were very much alike and they quickly became close friends. They both differed from others by their aristocratic personalities. It was Erhard Schmidt who directed Carathéodory to Göttingen. In Göttingen he became most popular among the members of the young mathematical generation. He received his PhD in 1904 and already less than one year later, in 1905, just at the end of his fifth year at the university, his *Habilitation*. This was and still might be a record at German universities and broke a rule suggested by nobody less than Hilbert. In 1908 he became a lecturer in Bonn. By 1909 he was a full professor at the Technical University in Hanover. One year later he changed to the Technical University in Breslau where Erhard

Constantin Carathéodory

Schmidt was employed at that time and in 1913 he became Felix Klein's successor in Göttingen.

When Frobenius died in 1917, the faculty at the University of Berlin suggested the following nominees for a successor:

1. Issai Schur, Constantin Carathéodory.
2. Hermann Weyl.

Carathéodory got the call and accepted. One of the reasons for his doing so was that Erhard Schmidt just one year before had replaced H.A. Schwarz at this university. Carathéodory started on October 1, 1918, at the University of Berlin exactly 12 months after Schmidt. On February 10, 1919, he was elected member of the Prussian Academy of Sciences. Everything seemed to be prepared for another flourishing period of mathematics in Berlin.

It was his fate and that of mathematics and of mathematicians in Berlin that he had to resign from his positions at the University of Berlin and the Academy only after 15 months, when his patriotism forced him to follow the call of his country, Greece. In September 1919 the Greek Prime Minister E. Venizelos, while at a peace conference in Paris, had invited Carathéodory to discuss educational problems of Greece with him. A few weeks later Carathéodory had submitted a plan to the Greek government for the foundation of a new, second university in Greece. The government had decided to found this university in the recently liberated city of Smyrna and had invited Carathéodory to organize it. He accepted and quit his positions in Berlin at the end of 1919. As a consequence he became honorary member of the Academy in Berlin in 1920.

Smyrna soon was conquered by the Turks again and Carathéodory had to flee in September 1922 (see [7], introduction). He was immediately appointed professor at the University of Athens and in 1923 also at the National Technical University of Athens. One year later he followed a call as the successor of Lindemann from the University of Munich. There he remained for the rest of his life, about 25 years. He became a member of the Bavarian Academy of Sciences and of many other academic societies. He especially appreciated his membership to the Papal Academy which was very exceptional.

Due to his knowledge and his elegance he was often asked to give speeches. He was the one who in 1936 handed over the first Field medals at the ICM in Oslo.

Once again he was involved with the organization of universities when, in 1930, the Greek government asked him to act as State Representative at the Universities of Athens and Thessaloniki. Besides his reports on the organization of a university he wrote papers on educational problems. His mathematical œuvre is both important and many-sided. He worked in different fields of mathematics, as well in pure as in applied areas. But contrary to many others he kept working in these fields all his life and not just for some periods. His main areas were function theory, real analysis and variational calculus. He also did work in geometry and partial differential equations. But, being an engineer after all, he also investigated problems in applied mathematics besides his studies in pure mathematics. His research in thermodynamics was noticed by well-known physicists of his time such as Planck, Born, and Sommerfeld. He also did work in mechanics, geometric optics and relativity theory. His great interest in history, especially of mathematics, led him to write historical bibliograhical papers.

His interest in classical complex analysis was stimulated by H.A. Schwarz in Berlin and also influenced by his personal contact with Landau. It was Carathéodory who proved the boundary correspondence under conformal mappings of Jordan domains. And he solved the problem for arbitrary simply connected domains by creating the theory of "prime ends". He has given an elementary deduction of what he has called the "Schwarz Lemma". In the theory of several complex variables he was looking for analogues of classical results from function theory of one complex variable. He surprised mathematicians with the fact that the ball in $\mathbf{C}^2$ is not holomorphically equivalent to the bidisc.

Stimulated by the results of Lebesgue on measure and integral he worked on real functions. His condition for sets to be Lebesgue-measurable is nowadays known and used everywhere. But he has in an axiomatic way created a more general measure theory leading to his theory of "Somen". Carathéodory is often referred to in connection with nonlinear differential equations the nonlinearity of which is often assumed to satisfy the "Carathéodory condition". Already in his PhD thesis Carathéodory was involved with variational calculus. Here he extended the solutions to discontinuous ones and considered isoperimetrical problems already.

His main merit in this theory is his observation of its close relation to the theory of partial differential equations. His viewpoint was quite different from that of his predecessors and thus he revolutionized this theory.

Carathéodory wrote remarkable books in all these different areas which were of great influence and some of which are still used nowadays (see [8] – [15], [17]). His papers, articles, speeches and obituaries have appeared everywhere (see [3], [16]). He was well-known, respected and liked throughout the mathematical community.

References

[1] *Behnke, H.*, Carathéodory's Leben und Wirken, The Greek Math. Soc., C. Carathéodory, Intern. Symp., Athens, Sept. 1973, Proceedings, Athens, 1974, 17–33.

[2] *Carathéodory, C.*, Autobiographische Notizen, Ges. Math. Schriften 5, 387–408.

[3] *Carathéodory, C.*, Zusammenstellung der Schriften C. Carathéodorys, Ges. Math. Schriften 5, 421–433.

[4] *Perron, O.*, Constantin Carathéodory, Jber. dt. Math.-Verein. **55** (1952), 39–51.

[5] *Schmidt, E.*, Constantin Carathéodory zum 70. Geburtstag, Forschungen und Fortschritte **19** (1943), 249–250.

[6] *Schmidt, E.*, Constantin Carathéodory zum 70. Geburtstag, in: Ges. Math. Schriften 5, 409–419.

[7] *Rassias, Th. (ed.)*, Constantin Carathéodory: an international tribute, 2 vols., World Scientific, Singapore, 1991.

[8] *Carathéodory, C.*, Analytische Geometrie, Vorlesung Sommersemester 1918, Göttingen 1918.

[9] *Carathéodory, C.*, Vorlesungen über reelle Funktionen, Teubner, Leipzig und Berlin, 1918; 2. Aufl. 1927; Nachdruck Chelsea, New York, 1948; 3. korr. Aufl. 1968.

[10] *Carathéodory, C.*, Conformal representation, Cambridge Univ. Press, Cambridge, 1932.

[11] *Carathéodory, C.*, Variationsrechnung und partielle Differentialgleichungen erster Ordnung, Teubner, Leipzig und Berlin, 1935; 2. Aufl. 1956; Engl. trans., Calculus of variations and partial differential equations of first order, Holden-Day, San Francisco, 1967.

[12] *Carathéodory, C.*, Geometrische Optik, Springer-Verlag, Berlin, 1937.

[13] *Carathéodory, C.*, Reelle Funktionen, Zahlen, Punktmengen, Funktionen, Teubner, Leipzig und Berlin, 1939.

[14] *Carathéodory, C.*, Elementare Theorie des Spiegelungsteleskops von B. Schmidt, Teubner, Leipzig und Berlin, 1940.

[15] *Carathéodory, C.*, Funktionentheorie, 2 vols., Birkhäuser, Basel, 1950.

[16] *Carathéodory, C.*, Ges. Math. Schriften 1–5, C.H. Beck'sche Verlagsbuchhandlung, München, 1954–1957.

[17] *Carathéodory, C.*, Maß und Integral und ihre Algebraisierung, Birkhäuser, Basel, Stuttgart, 1956.

[B] *Biermann, K.-R.*, Die Mathematik und ihre Dozenten an der Berliner Universität 1810–1933, Akademie Verlag, Berlin, 1998.

Richard von Mises

Hans Föllmer and Uwe Küchler

Richard von Mises was professor at the University of Berlin from 1920 to 1933. During these thirteen years he advanced, with great vigour, the field of applied mathematics and stochastics. His research contributions cover a most impressive range, and the impact of his ideas can be felt even today.

Richard Martin Edler von Mises was born on April 19, 1883, in Lemberg (Lwow). He studied mathematics, physics and engineering at the *Technische Hochschule* in Vienna where Emanuel Czuber, Theodor Schmidt and Emil Müller were among his teachers. He then worked as an assistant with Georg Hamel in Brünn (Brno). In 1908 he took his doctor's degree in Vienna with a thesis on a topic in engineering (*Ermittlung der Schwungmassen im Schubkurbelgetriebe*). In the same year he received his *Habilitation* as a lecturer for engineering and machine construction in Brünn, and in 1909 he became associate professor of applied mathematics at the University of Strasbourg. During the First World War he served in the Austro-Hungarian army, first as a pilot and then as a planner, constructor and lecturer for its new air force. Already in 1913 he had started to lecture on the design of airplanes. His lectures were published in 1916 as the book *Fluglehre* [12] which became widely recognized as a pioneering text for this new technology. In 1919 he was appointed to the new chair of hydro- and aerodynamics at the *Technische Hochschule* in Dresden.

In 1920 Richard von Mises accepted a call to the University of Berlin and became director of a new Institute of Applied Mathematics. In the following years he expanded the role of applied mathematics with great energy and efficiency, both in teaching and in research. Alexander Ostrowski said in a talk in 1965: "Only with the appointment of Richard von Mises to the University of Berlin did the first mathematically serious German school of applied mathematics with a broad sphere of influence came into existence" ([18]). In particular, von Mises developed a new curriculum in applied mathematics which started to attract many students. This included a *Mathematisches Praktikum* of six semesters, involving practical applications of mathematics in astronomy, geodesy and other technical areas. He emphasized that applied mathematics requires the same level of mathematical rigour as pure mathematics. Research topics for dissertations at the Institute of Applied Mathematics ranged from probability and statistics to numerical methods for differential equations and to applications of mathematics in elasticity and aerodynamics. Under his supervision seven mathematicians took their doctor's degree. Two received their *Habilitation*: Hilda Pollaczek-Geiringer and Stefan Bergmann. In addition to his vigorous contributions in education and research, von Mises was also very active as an organizer. In particular, he was founder and editor of the *Zeitschrift für Angewandte Mathematik und Mechanik* (ZAMM). In the introduction to the first volume of ZAMM, Richard von Mises defines the range of

Richard von Mises

applied mathematics from the point of view of an engineer (*was der Ingenieur an mathematischen Hilfsmitteln gebraucht, aus der Analysis und Geometrie, aus den verschieden verzweigten Teilen der Mechanik, aus der Thermodynamik und Elektrizitätslehre, aus der Wahrscheinlichkeitslehre und Statistik*). But he uses the term "engineer" in a very broad sense for anyone who works in a practical profession using the available scientific knowledge. At the same time, he stresses that the dividing line between pure and applied mathematics should not be taken too seriously and that it will vary over time, due to the increasing relevance of mathematics in many practical areas.

The extremely fruitful activities of Richard von Mises as a professor at the University of Berlin came to a sudden end in 1933. Being Jewish, he escaped the impending danger and became professor of mathematics at the University of Istanbul. During that time he worked increasingly on philosophical problems, and this led to an extensive treatise on positivistic philosophy [13]. In 1939 he moved to the United States. He was professor of mathematics at Harvard, and in 1944 he became Gordon McKay-Professor of Aerodynamics and Applied Mathematics. He received several honorary degrees and was elected as member of the American Academy of Sciences and as corresponding member of the Bavarian Academy of Science. In 1950 he was elected as a corresponding member of the German Academy of Sciences in the GDR. In a letter of September 15, 1950 [1], he writes that he feels obliged to decline this invitation even though he would be very pleased, in memory of his time as a professor in Berlin, to renew his ties to scientific life in Germany. But in the prevailing situation both in the United States and in Germany his acceptance might be viewed as a political statement, and he has made it a rule throughout his life to keep away from any political involvement. After a long illness Richard von Mises died on July 14, 1953, shortly after his 70th birthday.

In 1921, von Mises had asked Hilda Pollaczek-Geiringer, who had received her doctor's degree in Vienna with a thesis on Fourier series, to become his assistant at the Institute of Applied Mathematics. She was in charge of the *Mathematisches Praktikum* and shifted her research interests to applied mathematics and probability. After a somewhat controversial procedure ([18]) she received her *Habilitation* in 1927, as the first woman in Berlin and as the second in Germany (after Emmy Noether) in the field of mathematics. Like Richard von Mises and Stefan Bergmann (who became a professor in Tomsk), she was forced in 1933 to leave the University of Berlin. She joined von Mises in Istanbul and they married in 1943. After their move to the United States, Hilda Geiringer became professor of mathematics at Wheaton College in Massachusetts.

In his research, Richard von Mises covered an impressive range of topics. Shortly before his death he classified his contributions into eight groups: practical analysis, integral and differential equations, mechanics, hydro- and aerodynamics, constructive geometry, probability calculus, statistics and philosophy. To all these areas he has made important and lasting contributions. One example is his work in plasticity where he introduced a new invariant of the stress tensor which greatly simplified the computation of the strength of materials. This is now widely known among engineers as the "von Mises hypothesis". Another example is the Bernstein-von Mises theorem in the asymptotic analysis of functionals of the empirical distribution in statistics. Such functionals are now called von Mises functionals, and the basic idea of differentiating these functionals, a key tool in robust statistics, goes back to him.

Here we only focus on his contribution to the foundations of probability. After its appearance in 1919, his frequentist approach based on the notion of a *Kollektiv* generated a lot of interest, including severe criticism. In the fourties it started to disappear almost completely from the main stream development of probability theory. But in the sixties the ideas of Richard von Mises reemerged with lasting impact as the source of an important new development.

Richard von Mises was convinced that the concepts and methods of probability and statistics should be developed in a systematic manner from the notion of a "random sequence" or, as he called it, a *Kollektiv*. Such a collective describes a series of experiments whose outcomes vary in some space M. It is defined in [14] as a sequence $(x_1, x_2, \ldots)$ with values $x_i \in M$ such that the following property holds: Each subset A of M appears with some asymptotic relative frequency W_A, not only along the sequence itself but also along each subsequence which is formed without advance knowledge of the upcoming values (*ohne Berücksichtigung der Merkmalsunterschiede*). Thus, each collective induces a probability distribution W on the space M (*Erst das Kollektiv, dann die Wahrscheinlichkeit*), and this distribution is invariant under "place selection" (*Prinzip der Regellosigkeit oder des ausgeschlossenen Spielsystems*). Collectives can be combined to form new collectives, and von Mises defines the task of probability theory as "the study of the change of distributions implied by such transformations."

It was soon noticed, in particular by Felix Hausdorff, that there are serious problems with this notion. Apart from the measure theoretic problems which arise in the case of an uncountable space M, there was a fundamental flaw which appears already in the simplest case $M = \{0, 1\}$. Hausdorff pointed out that an unrestricted invariance under "place selection" implies that collectives with a nondegenerate distribution cannot exist. In a letter to Pólya in January 1920, he concluded that the von Mises approach is a total failure (*total verunglückt*). In the thirties there was a first systematic attempt by several mathematicians and philosophers, including Popper, Reichenbach, Copeland, Wald and Church, to develop a rigorous version of the von Mises approach and to reconcile it with Kolmogorov's axiomatic foundation of probability theory in terms of measure theory [6] which had appeared in 1933. This led to the notion of a "Bernoulli sequence" (*nachwirkungsfreie Folge*). Such a sequence has the correct relative frequencies along any subsequence which is formed by the following rule: Fix a word of some finite length with alphabet M, then take all terms of the original sequence which come right after an occurrence of this word. In the Kolmogorov approach, a stationary sequence of independent experiments is described by a "Bernoulli measure" on the space of all M-valued sequences, i.e., a product measure whose marginal distribution is given by the probability distribution of a single experiment. In modern terminology, the Bernoulli sequences are exactly the sequences such that the associated empirical process converges weakly to the Bernoulli measure. Due to the ergodic theorem, the set of such sequences has Bernoulli measure 1. Thus, the formulation of the von Mises program in terms of Bernoulli sequences is certainly consistent with the framework of Kolmogorov. Abraham Wald observed in 1936 that the same is true if the notion of a collective is formulated in terms of a countable class of place selection rules where the decision to include the next term is defined as a function of the preceding values. In particular, this applies to the class of computable selection rules introduced by Church in 1940. But in 1939 came what many perceived as the *coup de grâce* for the von Mises approach. Jean Ville showed in his *Etude critique de la notion du collectif* that for any countable class of place selection rules there is a collective which violates the law of the iterated logarithm. This law was proved in 1924 by Chintchin, and since then it was viewed as an essential feature of "random behaviour." Richard von Mises reacted to this news with the comment "*J'accepte ce théorème, mais je n'y vois pas une objection.*" But most probabilists felt that Ville's result had revealed a basic deficiency in the notion of a collective. Moreover, the measure theoretic setting of Kolmogorov provided a more elegant and efficient methodology than the direct manipulation of collectives proposed by von Mises; see the debate in 1940 at Dartmouth College between von Mises and J.L. Doob [15]. Thus, interest in the von Mises approach started to fade in the fourties, and the emphasis shifted almost completely to Kolmogorov's axiomatic framework. This is reflected in the following comment of Ostrowski right after his praise of von Mises quoted above: "Because of his dynamic personality his occasionally major blunders were somehow tolerated. One has even forgiven him his theory of probability." But such a dismissal failed to appreciate the fact that von Mises had posed a very fundamental problem which was left unsolved in Kolmogorov's axiomatic setting: How do we see that a given sequence is "random"?

It was Kolmogorov himself who clearly saw the need to return to von Mises's original questions. In 1963 he wrote "that the basis for the applicability of the results of the mathematical theory of probability to real 'random phenomena' must depend on some form of the *frequency concept of probability*, the unavoidable nature of which has been established by von Mises in a spirited manner" [7]. Kolmogorov introduced the idea of defining a "random sequence" in terms of how difficult it is to describe it. He turned this idea into a precise method based on algorithmic complexity and information theory ([7], [8]). This led to a strong revival of the original ideas of von Mises and to their implementation in algorithmic form. The idea that a random sequence should be such that it exludes the possibility of a long run gain by means of some gambling system (*Prinzip vom ausgeschlossenen Spielsystem*) became a key theme in the algorithmic approach of C. P. Schnorr [17]. The related idea that a random sequence should pass any statistical test based on one of the "almost sure laws" of probability theory (including the law of the iterated logarithm) was made precise by P. Martin-Löf [11]. Since the sixties the mathematical theory of random sequences has become a vital area of mathematical research at the interface of probability, information theory, algorithmic complexity, and logic (see, e.g., [4], [10]). It is widely agreed that Richard von Mises should be seen as one of its founding fathers.

References

[1] Archiv der Berlin-Brandenburgischen Akademie der Wissenschaften, Bestand Akademieleitung, Personalia, Nr. 315.

[2] *Bernhardt, H.*, Zum Leben und Wirken des Mathematikers Richard von Mises, NTM-Schriftenreihe für Geschichte der Naturwissenschaften, Technik und Medizin **16** (2) (1979), 40–49.

[3] *Bernhardt, H.*, Richard von Mises und sein Beitrag zur Grundlegung der Wahrscheinlichkeitsrechnung im 20. Jahrhundert, Diss. B, Humboldt-Universität, Berlin, 1983.

[4] *Chaitin, G.J.*, Algorithmic Information Theory, Cambridge Univ. Press, Cambridge, 1987.

[5] *Cramer, H.*, Richard von Mises' Work on Probability and Statistics, Ann. Math. Stat. **24** (1953), 657–662.

[6] *Kolmogorov, A.N.*, Grundbegriffe der Wahrscheinlichkeitsrechnung, Ergeb. Math. 2, No. 3, Springer-Verlag, Berlin, 1933.

[7] *Kolmogorov, A.N.*, On tables of random numbers, Sankhya Ser. A **25** (1963), 369–376.

[8] *Kolmogorov, A.N.*, Three approaches to the quantitative definition of information, Problems of Information Tansmission **1** (1) (1965), 1–7.

[9] *Küchler, U.*, Richard v. Mises (Laudatio), Statistics **14** (1983), 507–508.

[10] *van Lambalgen, M.*, Von Mises' definition of random sequences reconsidered, J. Symbolic Logic **5** (1987), 725–750.

[11] *Martin-Löf, P.*, The definition of random sequences, Inform. Control B (1966), 602–619.

[12] *von Mises, R.*, Fluglehre, 5. Aufl., Berlin, 1936.

[13] *von Mises, R.*, Kleines Lehrbuch des Positivismus, The Hague, 1939.

[14] *von Mises, R.*, Grundlagen der Wahrscheinlichkeitstheorie, Math. Z. **5** (1919), 52–99.

[15] *von Mises, R.*, Selected Papers, vol. 2, American Mathematical Society, 1964.

[16] *von Mises, R.*, Wahrscheinlichkeit, Statistik und Wahrheit, Springer-Verlag, Wien, 1951.

[17] *Schnorr, C.P.*, Zufälligkeit und Wahrscheinlichkeit, Eine algorithmische Begründung der Wahrscheinlichkeitstheorie, Springer-Verlag, Lecture Notes Math. 218, 1970.

[18] *Siegmund-Schultze, R.*, Hilda Geiringer-von Mises, Charlier Series, Ideology, and the Human Side of the Emancipation of Applied Mathematics at the University of Berlin during the 1920s, Hist. Math. **20** (3) (1993), 364–381.

Einstein in Berlin

David E. Rowe

On 12 November 1913, Kaiser Wilhelm II officially approved the appointment of Albert Einstein to membership in the Prussian Academy of Sciences. The following month Einstein accepted an offer, arranged by Max Planck and Walther Nernst, to become director of the newly established Kaiser Wilhelm Institute for Physics in Berlin. Thus began the most dramatic period in a singularly eventful career. Aside from a brief stay in Prague, Einstein had spent his entire adult life in Switzerland, where since 1912 he was professor of physics at the ETH in Zürich. Although born and raised in southern Germany, he had given up German citizenship in 1896 to become a Swiss citizen five years later. Accepting the offers from Berlin thus meant not only leaving these familiar surroundings but also giving them up to enter the hectic scientific life and turbulent political atmosphere of late Wilhelmian Germany. Few, least of all someone like Einstein, could have imagined the precarious events that lay just ahead. Einstein's arrival nearly coincided with the outbreak of the war that eventually brought about the fall of imperial Germany; his departure for the United States in December 1932 came as the Weimar Republic found itself on the brink of collapse. He broke his own official ties with Berlin by resigning from the Prussian Academy on 28 March 1933, eight days after a band of Nazis raided his unoccupied summer home in Caputh.

The Berlin years were as tumultuous for Einstein scientifically as they would be for Germany politically. Prior to his arrival he was seen by many as the rising star whose ideas largely set the tone for a fundamentally new reorientation toward the problems of mainstream physics. At the 1911 Solvay Conference, Einstein's presentation on "The Current Status of the Problem of Specific Heats" had made a deep impression on Walther Nernst, despite the scepticism Einstein had expressed regarding the status of Nernst's chief claim to fame, the third law of thermodynamics. By attracting him to the Prussian capital, both Nernst and Planck clearly hoped that the new Kaiser Wilhelm Institute for Physics would emerge as a major center for research in the microphysics of matter and radiation. Presumably Einstein himself harbored similar hopes.

During the two years prior to his arrival, however, Einstein had become increasingly absorbed in an effort to develop a relativistic theory of gravitation (see [14] and [21]; for a preview of recent historical work done by the "Berlin Relativity Group," see [18].) Practically from the moment he had returned to Zürich from Prague in 1912, he began collaborating with his long-time friend, the mathematician Marcel Grossmann, in a lonely struggle to develop such a theory. It was Grossmann who first pointed out to Einstein that the tools required to express properties that were generally covariant (as opposed to those covariant under a restricted group of transformations, such as the Lorentz transformations of special relativity theory) could be found in the theory of Riemannian spaces as well

Albert Einstein

as the tensor analysis developed by Gregorio Ricci-Curbastro. (For mathematical developments connected with tensor analysis, see [17].) With Grossmann's help Einstein soon mastered the basic techniques of the Ricci calculus, the first leg in a long intellectual journey that took him further and further away from the main avenues of research then being explored and travelled by his fellow physicists.

The young Einstein had always scorned what he called "formal approaches" to physical problems, an orientation he found typical among mathematicians (see [16]). Thus, Minkowski's space-time geometry for special relativity initially struck him as "superfluous learnedness," ([15], p. 152) and he later remarked to Felix Klein about the latter's treatment of the mathematical foundations of relativity theory: "it seems to me that you highly overrate the value of formal points of view. These may be valuable when an *already found* truth needs to be formulated in a final form, but they fail almost always as heuristic aids" ([15], p. 325). In the course of his work on general relativity theory, however, Einstein's attitude with respect to the utility of mathematics gradually underwent a striking change. After 1915, when he made the first major breakthrough by finding a system of generally covariant gravitational field equations, and even more strikingly after 1921, the year in which several new formalisms were presented for unifying gravity and electromagnetism, Einstein became increasingly convinced that only higher mathematics could unlock nature's deeper secrets. (For a detailed account of these developments, see [22].)

As Einstein's research interests drifted away from the mainstream concerns of the physics community, he entered a realm of macrophysics whose practitioners included astronomers, cosmologists, differential geometers, and other mathematicians, but relatively few physicists. Initially, German astronomers evinced little interest in general relativity theory, no doubt partly because its results were perceived as having little apparent import for the field. Moreover, the three effects Einstein was able eventually to deduce by using it – the correction to the precession of Mercury's perihelion motion, deflection of light in the presence of a gravitational field, and gravitational redshift – could be regarded in the first case as a matter of fine tuning Newton's theory and, in the other two instances, as involving minute measurements that made it difficult to detect the predicted phenomena. Nevertheless, two astronomers – Erwin Freundlich and Karl Schwarzschild – did take a keen and early interest in Einstein's theory, being among the first to undertake investigations aimed at confirming its empirical predictions. In 1911, soon after Einstein published his first paper predicting the bending of light in the vicinity of a gravitational field, Freundlich opened a correspondence with him about the possibility of detecting this effect during a solar eclipse (see [12], pp. 147–150). With Einstein's help, Freundlich succeeded in obtaining the necessary financial support for this project, so that in the summer of 1914 he headed a German expedition to the Crimea whose purpose was to measure the deflection of light rays during the solar eclipse of August 21. Due to the outbreak of the war, however, this mission had to be abandoned. Around the time Freundlich was readying his optical instruments for this ill-fated venture, Schwarzschild was in Potsdam studying bands of light emanating from the sun in an effort to detect gravitational red shift. Although these results proved negative, Schwarzschild remained enthusiastic about Einstein's theory, writing to Sommerfeld that "I'm surprised by my own relative indifference as to the empirical confirmation of general relativity. I'm just satisfied to go walking in such a beautiful *Gedankengebäude*" (quoted in [1], p. 68).

Both Freundlich and Schwarzschild had earlier been associated with the dynamic Göttingen atmosphere surrounding Klein and Hilbert before they joined forces with Einstein in Berlin. In the summer of 1915 Einstein made his first direct encounter with this Göttingen milieu when he delivered a series of six two-hour lectures there on general relativity theory. (He had been invited to Göttingen once before by Hilbert, in 1912, but declined that invitation; Albert Einstein to David Hilbert, 4 October 1912, in [8], pp. 321, 322). Einstein was pleased with the reception he was accorded, expressing particular pleasure with Hilbert's reaction. This visit also undoubtedly caused Hilbert to reintensify his own efforts to incorporate gravitational effects into Gustav Mie's electromagnetic theory of matter, an approach that at the same time meant generalizing the flat space-time geometry his friend Minkowski had publicized shortly before his unexpected death in 1909. In particular, Hilbert was initially persuaded by Einstein's then current belief that no generally covariant theory of gravitation could be obtained that preserved causality. Moreover, as recently publicized by a team of researchers associated with the

Max Planck Institut für Wissenschaftsgeschichte in Berlin, Hilbert continued to hold fast to this misconception right up until November 1915 (see [3]).

That autumn Einstein and Hilbert had been corresponding regularly, and by early November both were close to achieving major breakthroughs (see [4], [15], pp. 250–261). On 18 November Einstein announced that his theory could account for the mysterious anomaly in the precession of Mercury's orbit, a phenomenon that had been haunting celestial mechanics ever since LeVerrier first called attention to it in 1859 (see [19] and [5]). Afterward, Simon Newcomb had quantified this effect precisely, pointing out that the perihelion of Mercury's orbit advances at a rate of 575 seconds of arc per century, whereas Newton's theory predicted a precession of only $532''$ per century, leaving $43''$ unaccounted for. Numerous attempts had subsequently been made to explain this discrepancy, but none of them were the least bit convincing. Einstein's result – assuming his theory was indeed sound – thus offered an astounding new insight by showing that this recalcitrant problem could *never be* resolved on the basis of Newtonian gravitation alone since the missing $43''$ was a purely relativistic effect! "For a few days," he later wrote Paul Ehrenfest, "I was beside myself with excitement" (quoted in [15], p. 253).

Only two days after Einstein's announcement, Hilbert presented the first of two notes on "The Foundations of Physics" to the Göttingen Scientific Society. Applying variational principles to a Lagrangian "world function," an approach based on Gustav Mie's electromagnetic theory of matter, Hilbert derived a series of results for a unified theory of gravity and electromagnetism. Five days later, on 25 November, Einstein submitted the last of his four notes on the gravitational field equations to the Berlin Academy ([6]). This note contains the derivation of the fundamental equations:

$$R_{\mu\nu} = -\kappa\Big(T_{\mu\nu} - \frac{1}{2}g_{\mu\nu}T\Big).$$

In recent decades, historians have struggled to sort out an apparent priority issue stemming from the fact that Hilbert's paper was *dated* five days earlier than Einstein's (see, for example, [4], [15], pp. 257–261). Since Hilbert's paper contains a different derivation of a closely related form of the above field equations, the question naturally arises whether he could justifiably claim a share of credit for this important finding. It should be emphasized, however, that neither Hilbert nor any of his contemporaries ever questioned Einstein's priority with regard to the gravitational field equations. Thanks to diligent archival work by Leo Corry, this "priority issue" as to whether Hilbert anticipated Einstein's discovery of generally covariant field equations can finally be put to rest (see [3]).

As it turns out, the paper Hilbert presented on 20 November differed *significantly* from the version that actually appeared in print on 31 March 1916. In the meantime, Hilbert had apparently grasped that causality could, in fact, be reconciled with a generally covariant gravitational theory, an insight he apparently only reached after having read Einstein's November articles. What is more, at the time Hilbert initially submitted his paper it contained no explicit gravitational field equations at all! He simply added these into the text when he received page proofs in December along with a brief claim that they follow "easily without calculation" from the formal properties of the metric and Ricci tensors (the $g_{\mu\nu}$ and $R_{\mu\nu}$ in the equations above). While this might sound like a clearcut case of plagiarism, one should keep in mind that Hilbert's intensely interactive research style often necessitated that he make use of other mathematicians' insights. Not surprisingly, this mode of work sometimes led him to appropriate foreign ideas – a phenomenon that came to be known as "nostrification" in Göttingen circles – without necessarily realizing the intellectual debts he may have owed others. We will presumably never know whether charity, malice, or simple ambition played a role when Hilbert decided to alter his paper from 20 November, but we can at least be sure that he was in no position whatsoever to challenge Einstein's priority with respect to the field equations.

It also remains unclear whether Einstein knew anything about these late alterations, although we do know that he was none too happy about Hilbert's behavior that December and wrote to let him know so:

> There has been a certain resentment between us, the cause of which I do not wish to analyze any further. I have fought against the feeling of bitterness associated with it, and with complete success. I again think of you with undiminished kindness and I ask you to attempt the same with me. It is objectively a pity if two guys who have somewhat liberated themselves from this shabby world are not giving pleasure to each other. (Einstein to Hilbert, 20 December 1915, quoted in [3], p. 1273.)

These intense efforts left Einstein in a state of mixed delight and exhaustion, but in a very real sense his work on general relativity had just begun. Unlike the situation earlier, however, his efforts would from now on be supported by numerous collaborators and other interested parties. Indeed, the next significant step forward was taken by Schwarzschild, who realized that Einstein's calculation of the displacement in Mercury's perihelion was based on a rather crude method of making approximations. In order to rectify this, Schwarzschild published a paper in January 1916 that gave the first solution to the gravitational field equations by taking the case of a single mass point. In a follow-up paper, he then handled the case of an incompressible liquid sphere, which led to some remarkable findings (see [10]). At points far removed from the sphere, the gravitational field resembled that found in the case of a single mass point. At points within the sphere, however, the gravitational field forms a spherical geometry whose radius of curvature depends on the density of the fluid, but which is always greater than the radius of the sphere. Schwarzschild wrote these papers, which were to play a major role for decades to come, while stationed on the eastern front. They were presented by Einstein to the Prussian Academy in early 1916. Soon thereafter Schwarzschild contracted a skin disease that took his life on 11 May, 1916. The following month

Einstein spoke of his achievements at a commemorative meeting held in his honor in the academy (see [7]).

While certain that he was now on the right track, Einstein was not one to bask in the glow of a triumph for very long. Probably more than anyone, he recognized that many pressing theoretical problems remained to be tackled and that the next few years would be of decisive importance for the reception of his theory. In approaching the tasks that lay ahead, he took a remarkably pragmatic, open-ended view. For him, the laws of physics had to be generally covariant, but he was willing to tinker with almost everything else, including the form of the field equations. This attitude stood in sharp contrast with Hilbert's rather rigid position. Unlike Einstein, Hilbert derived his results from a variational principle applied to a special Lagrangian based on Mie's electromagnetic theory of matter. His aim was to show that Einstein's gravitational theory could be combined with Mie's electrodynamics, and that in a certain sense the latter was subsumed by the former. Vladimir Vizgin has aptly described Hilbert's approach as "doubly reductionistic – matter was reduced to the electromagnetic field, and the equations of the electromagnetic field were deduced from the equations of the gravitational field" ([22], p. 61).

While Hilbert's effort may be regarded as the first unified field theory, it failed to find many adherents. The ever-optimistic Hilbert nevertheless saw his theory as a major triumph for the axiomatic method, claiming that by means of "essentially only two simple axioms" (in the paper submitted on 20 November 1915 he had to employ a third axiom; see [3], p. 1272) he could deduce "a new system of fundamental equations for physics of ideal beauty ... and which contain *simultaneously* the solution to the problems of Einstein and Mie" ([13], p. 395). Furthermore, Hilbert expressed his conviction "that the most minute, till now hidden processes within the atom will become clarified through the herein exhibited fundamental equations, and that it must be possible in general to refer all physical constants back to mathematical constants..." ([13], p. 407).

Einstein took a very different view of Hilbert's achievement. He recognized the importance of deriving the field equations from an appropriate variational principle, but he also strongly felt that nothing was to be gained by linking his theory of gravitation with Mie's field-theoretic approach to matter. Writing to Hermann Weyl, he confessed that "to me Hilbert's *Ansatz* about matter appears to be childish, just like an infant who is unaware of the pitfalls of the real world..." (A. Einstein to H. Weyl, 23 November 1916, quoted in [20], p. 200). But Einstein was also highly critical of Weyl's more sophisticated unified field theory as first presented in his 1918 paper *Gravitation und Elektrizität* ([23]). (Weyl later elaborated his theory in the third edition of his famous book *Raum-Zeit-Materie*). Weyl's approach, which utilized Levi-Civita's new technique of infinitesimal parallel displacements, was far more flexible than Hilbert's. Nevertheless, Einstein argued that in Weyl's theory the rate of clocks and the length of rods depend on their prehistory, which cannot be reconciled with the fact that we observe sharp spectral lines.

Einstein evinced real interest in early efforts to unify gravity and electromagnetism, but his own main concerns between 1916 and 1921 centered on issues involving the status of energy conservation in general relativity as well as the cosmological implications of the field equations. He also began to contemplate the deeper implications of quantum theory during this period. In his 1917 paper, "Cosmological Considerations on the General Theory of Relativity," Einstein introduced the cosmological constant λ while setting forth his first published reflections on the global space-time geometry of a static universe. This meant modifying the original field equations by adding the cosmological term $\lambda g_{\mu\nu}$ in order to ensure static solutions (in 1929 Hubble unveiled evidence for an expanding universe, causing Einstein to abandon the cosmological term two years later).

Before 1921, Weyl's was the only fully elaborated unified field theory going, but one year later a plethora of competitors had made their appearance on the scene. A central goal of this research was to build an integrated theory that could potentially make room for quantum phenomena. Thus, Arthur Eddington generalized Weyl's theory by utilizing the coefficients of the affine connection Γ^i_{kl} arising from infinitesimal parallel displacement. Around the same time, Theodor Kaluza presented the first 5-dimensional unified field theory, an approach later taken up by the Swedish mathematician, Oskar Klein. Even Weyl, partly in reaction to Einstein's criticisms, set forth a different version of his original theory, which he elaborated in the fifth and final edition of *Raum-Zeit-Materie*, published in 1923. Thereafter, Weyl chose to quit the arena of unified field theories, after which time Einstein himself emerged as the leading theoretician, or at least as the one with the greatest staying power.

As he gradually became estranged by the radically new underpinnings that Bohr, Heisenberg, Born, et al., had given to quantum physics, Einstein began focusing serious attention on a variety of schemes to unify gravity and electromagnetism, even ignoring their potential for an eventual reconciliation with quantum theory. Working closely with collaborators like Cornelius Lanczos, Hermann Müntz, Walther Mayer, but especially with the Russian-born mathematician Jakob Grommer, his assistant in Berlin for some ten years, he eventually tried his hand at nearly every technique and model available. (On Einstein's collaborators, see [15], pp. 483–497.) Between 1929–1932 he carried on a lengthy correspondence with Elie Cartan, who had already given a classification of generalized Riemannian spaces that admit an affine connection back in 1922–23 (see [2]). By the time he left Berlin, enthusiasm for unified field theories had all but evaporated. Still, Einstein persisted in the hunt right up until his death in 1955.

This brief sketch should at least convey one clear point, namely that any attempt to contextualize Einstein's activities during his Berlin period must take into account his many ties outside the physics community proper. In particular, his relationships with mathematicians took on a growing importance during these years in which Einstein became increasingly enamored with the full cornucopia of potential mathematical formalisms in his search for a satisfactory unified field theory. Convinced that the new quantum mechanics would eventually dissolve into a modern, unified field physics, and that the latter could best be secured by developing the proper framework for an integration of gravitational and electromagnetic forces, Einstein turned his back on the younger generation so naturally attracted to the excitement and challenge of work in quantum physics. By the time he left Berlin and settled into the quiet seclusion of Princeton's Institute for Advanced Study, the physicist Einstein had become a recluse to the community that had earlier followed his ideas with such intense fascination.

Acknowledgements

The present account relies heavily on the excellent scientific biography by Abraham Pais [15]. For further biographical details, see [11]. Much of the correspondence referred to above can be found in [9]. The author wishes to thank Leo Corry, Tilman Sauer, and especially Michel Janssen for their helpful criticisms of an earlier draft of the present essay.

References

[1] *Blumenthal, O.*, Karl Schwarzschild, Jber. dt. Math.-Verein. **26** (1916), 56–75.

[2] *Debever, R. (ed.),* Elie Cartan–Albert Einstein, Letters on Absolute Parallelism, 1929–1932, Princeton Univ. Press, Princeton, 1979.

[3] *Corry, L., Renn, J., Stachel, J.*, Belated Decision in the Hilbert-Einstein Priority Dispute, Science **278** (1997), 1270–1273.

[4] *Earman, J., Glymour, C.*, Einstein and Hilbert: Two Months in the History of General Relativity, Arch. Hist. Exact Sci. **19** (1978), 291–308.

[5] *Earman, J., Janssen, M.*, Einstein's Explanation of the Motion of Mercury's Perihelion, in: Earman, J., Janssen, M., Norton, J.D. (eds.), The Attraction of Gravitation (Einstein Studies, vol. 5), Birkhäuser, Boston, 1993, 129–172.

[6] *Einstein, A.*, Die Feldgleichungen der Gravitation, Sitzungsberichte der Königlichen Preußischen Akademie der Wissenschaften (1915), 844–847.

[7] *Einstein, A.*, Gedächtnisrede des Hrn. Einstein auf Karl Schwarzschild, Sitzungsberichte der Königlichen Preußischen Akademie der Wissenschaften (1916), 768–770.

[8] The Collected Papers of Albert Einstein, vol. 5, Engl. trans. Anna Beck, Princeton University Press, Princeton, 1995.

[9] The Collected Papers of Albert Einstein, The Berlin years: Correspondence, 1914–1918, vol. 8, Schulmann, R., Kox, A.J., Janssen, M., Illy, J., Princeton University Press, Princeton, 1998.

[10] *Eisenstaedt, J.*, The Early Interpretation of the Schwarzschild Solution, in: Howard, D., Stachel, J. (eds.), Einstein and the History of General Relativity (Einstein Studies, vol. 1), Birkhäuser, Boston, 1989, 213–233.

[11] *Fölsing, A.*, Albert Einstein: A Biography, Viking, New York, 1997.

[12] *Hentschel, K.*, Erwin Finlay Freundlich and Testing Einstein's Theory of Relativity, Arch. Hist. Exact Sci. **47** (1994), 143–201.

[13] *Hilbert, D.*, Die Grundlagen der Physik (Erste Mitteilung), Nachrichten der Königlichen Gesellschaft der Wissenschaften zu Göttingen, Mathematisch-Physikalische Klasse (1915), 395–407.

[14] *Norton, J.*, How Einstein Found his Field Equations, in: Howard, D., Stachel, J. (eds.), Einstein and the History of General Relativity (Einstein Studies, vol. 1), Birkhäuser, Boston, 1989, 101–159.

[15] *Pais, A.*, Subtle is the Lord..., The Science and the Life of Albert Einstein, Clarendon Press, Oxford, 1982.

[16] *Pyenson, L.*, The Young Einstein, Adam Hilger, London, 1986.

[17] *Reich, K.*, Die Entwicklung des Tensorkalküls, Vom absoluten Differentialkalkül zur Relativitätstheorie, Science Networks, vol. 11, Birkhäuser, Basel, 1992.

[18] *Renn, J., Sauer, T.*, Einstein's Züricher Notizbuch, Physikalische Blätter **52** (1996), 865–872.

[19] *Roseveare, N.T.*, Mercury's Perihelion from Leverrier to Einstein, Clarendon Press, Oxford, 1982.

[20] *Seelig, C.*, Albert Einstein, Europa Verlag, Zürich, 1954.

[21] *Stachel, J.*, Einstein's Search for General Covariance, 1912–1915 in: Howard, D., Stachel, J. (eds.), Einstein and the History of General Relativity (Einstein Studies, vol. 1), Birkhäuser, Boston, 1989, 63–100.

[22] *Vizgin, V.P.*, Unified Field Theories in the First Third of the 20th Century, Science Networks, vol. 13, Birkhäuser, Basel, 1994.

[23] *Weyl, H.*, Gravitation und Elektrizität, Sitzungsberichte der Königlichen Preußischen Akademie der Wissenschaften (1918), 465–480; reprinted in Engl. trans. in: The Principle of Relativity, A Collection of Original Memoirs by H.A. Lorentz, A. Einstein, H. Minkowski, and H. Weyl, Dover, New York, 1952.

The Nazi era: the Berlin way of politicizing mathematics

Norbert Schappacher

The first impact of the Nazi regime on mathematical life, occurring essentially between 1933 and 1937, took the form of a wave of dismissals of Jewish or politically suspect civil servants. It affected, overall, about 30 per cent of all mathematicians holding positions at German universities. These dismissals had nothing to do with a systematic policy for science, rather they proceeded according to various laws and decrees which concerned all civil servants alike. The effect on individual institutes depended crucially on local circumstances – see [6], section 3.1, for details.

Among the Berlin mathematicians, 50-year-old Richard von Mises was the first to emigrate. He went to Istanbul at the end of 1933, where he was joined by his assistant and future second wife Hilda Pollaczek-Geiringer who had been dismissed from her position in the summer of 1933. Formally, von Mises resigned of his own free will – he was exempt from the racial clause of the first Nazi law about civil servants because he had already been a civil servant before August 1914. Having seen what the "New Germany" of the Nazis was like, he preferred to anticipate later, stricter laws, which would indeed have cost him his job and citizenship in the autumn of 1935.

The Jewish algebraist Issai Schur – who, like von Mises, had been a civil servant already before World War I – was temporarily put on leave in the summer of 1933, and even though this was revoked in October 1933 he would no longer teach the large classes which for years had been a must even for students who were not primarily into pure mathematics. In August 1935, aged 60, he gave in to the mounting pressure and submitted the request that he become emeritus. He emigrated to Palestine in 1939 where he died two years later.

Other Berlin mathematicians who lost their jobs or the right to teach university courses during the first years of Nazi rule included A. Brauer, St. Bergmann, as well as A. Barneck and E. Jacobsthal at the Technical University.

The most tragic case within the small community of Berlin mathematicians was that of the group and number theorist Robert Remak [5] whose marvellous regulator bounds are probably better appreciated by today's number theorists than they were in his time, and whose occasional papers on a general mathematical theory of economy have barely begun to be read. Remak had no position, only the right to lecture at the university. Even to obtain this so-called *Habilitation* had been a long and difficult process. His applications had been rejected several times. The Remak family had something of a tradition in slowly overcoming administrative hurdles at the University of Berlin. The mathematician's parental grandfather and namesake Robert Remak had been the first non-converted Jew to obtain, in 1847, his *Habilitation* at the Berlin medical faculty. A. v. Humboldt had lent a helping hand in his case. In the course of the first racial law, the mathematician Robert Remak lost his right to teach courses in September 1933 – altogether, he had held this right for only four years. He continued to live and do research in Berlin. After the so-called "Crystal Night," in November 1938, he was arrested and locked into the Sachsenhausen concentration camp for eight and a half weeks. During this time his (non-Jewish) wife tried to organize a place of refuge in case of his release. This resulted in his being allowed to go to Amsterdam, which he did in April 1939. But in May 1940 the German occupation caught up with him there. He was again arrested, brought to Auschwitz, where he was put to death on some unknown day after 1942.

The dismissals were impersonal, almost Kafkaesque, in that they manifested themselves as formal applications of general laws, executed by the same efficient administrative machine that had once been the pride of the Prussian state. An analysis of the application of the first law (from 7 April 1933) even shows that one of its aims was precisely to ward off the misconception, common in particular among the storm troopers (SA), of Hitler's seize of power as a revolution. Thus the temporary leave imposed on Schur in the summer of 1933 and his subsequent ban from large classes, are typical examples of the 1933 ministerial policy which tried to avoid Nazi student boycotts by withdrawing potential targets of such militant actions from major lecture courses. It also illustrates the pincer movement operated by the Nazi regime against its enemies: through street gangs on the one hand, and an allegedly respectable Prussian administration on the other.

But the Nazi-"revolutionaries" did not aquiesce, and would even find opportunities for direct action. This second kind of impact of Naziism on science was very different from the dismissals: much more ideological with direct attacks on scientific questions, or at least questions of scientific teaching, and more personal, without legal attire. Although there had been militant student actions against Jewish or politically disliked mathematicians before in several places (for instance, against Otto Blumenthal in Aachen), the real beginning of this second impact of Naziism on mathematics was marked by the boycott of Edmund Landau's lecture in Göttingen on 2 November 1933. This event was to have repercussions in the following years on the German mathematical community as a whole, and on mathematics in Berlin in particular.

The number theorist Edmund Landau was himself a Berliner, whom Klein and Hilbert had brought to Göttingen as Minkowski's successor in 1908. In the 1920s he had been among the co-founders and one of the first professors teaching at the Hebrew University in Jerusalem. During the spring term 1933, he was subjected to a similar treatment as Schur: his beginners' calculus class had to be delivered by his (Nazi) assistant Werner Weber, based on Landau's notes. Only at the beginning of the winter term, on November 2, 1933, did Landau try again to lecture himself; he was promptly stopped by a well-organized boycott, led by the extremely gifted young student Oswald Teichmüller, a militant member of the party and the SA, whose mathematical talent was matched only by his political fanaticism. Upon Landau's request, Teichmüller wrote up the "reasons" for this

Edmund Landau

boycott in an abominable letter to Landau which mixes a peculiar deference with sarcastic insults. The basic "reason" given by Teichmüller for the boycott was the alleged racial incompatibility between teacher and students: the attempt on the part of the student to adopt a presentation of calculus by a "teacher of a different race" would lead, Teichmüller claimed, to "intellectual degeneration" [7]. As a consequence of this boycott, Landau resigned from his chair, gave up his house in Göttingen, and moved to Berlin where he died on 19 February 1938. His last years were gloomy, illuminated only by a few mathematical trips to Holland and England.

Shortly after the Landau boycott, the Berlin mathematician Ludwig Bieberbach echoed Teichmüller's letter, and congratulated publicly the Göttingen students for their "manly" action against Landau – an action which he portrayed at the same time as a kind of biological necessity. During that winter term of 1933/34, Bieberbach was teaching a course entitled "Great German mathematicians, a race-theoretic approach" where he supposedly explained what was subsequently published in three papers: a classification of styles of mathematical production based on E.R. Jaensch's psychological typology of apperception (which was popular at the time, and was itself not originally based on racial ideas, although Jaensch joined the new trend later in the 1930s). This pseudo-scientific theory went well beyond a simple separation of "Aryan" from "Jewish" mathematics, but contained finer distinctions which of course brought inevitable problems of fitting certain

Ludwig Bieberbach

mathematicians into the right slot – Hilbert, for instance, was notoriously hard to accommodate in Bieberbach's theory.

It is one thing to reflect, as Bieberbach did, for instance on the relative pedagogical merit of different ways to introduce π in a calculus class: geometrically via the circle, or in Landau's way via the zeroes of the cosine, with this function being defined by its power series. And it is quite a different matter to use such reflections as a basis for advocating the forced removal of a distinguished colleague from teaching. Bieberbach's behaviour came all the more as a shock as nothing in his previous biography seemed to prepare one for it: Not only was he respected as a serious and deep mathematician (his work in geometric function theory actually continued to orient research well beyond his death), but no one had ever observed antisemitic tendencies in Bieberbach before the summer of 1933. He collaborated very well with Issai Schur for more than a decade; they even had a joint publication in 1928; but in 1935 it was Bieberbach more than anybody else who pushed Schur into early retirement. Bieberbach backed Remak's second attempt to get his *Habilitation* in 1923, against distinguished colleagues such as Max Planck. With Landau himself, Bieberbach had a mathematical correspondence soon after World War I, when Landau was going through a paper of Bieberbach's in a Göttingen seminar. In this correspondence, Bieberbach shows respect and admiration for Landau far beyond usual formulas of politeness. (The two halves of the correspondence are kept today at the Hebrew University in Jerusalem, resp. the

Göttingen Archives.) As far as his politics was concerned, Bieberbach was generally considered a loyal republican during the Weimar Republic – which made him an exception within the caste of German university professors who tended in general to cultivate monarchist ideals.

Yet there are elements of continuity in Bieberbach's attitudes, across the 1920s and 1930s: his vanity and love for honorary posts on the personal side, and his point of view on "nationalism vs. internationalism" in mathematics (as he himself put it in 1934) in the domain of mathematics policy.

His vanity was nicely recorded by Albert Einstein in 1919: "Herrn Bieberbach's love and admiration for himself and his muse is most delightful," he wrote to Born. In 1933, Bieberbach – who had been unfit for service in World War I – took part in a big SA march from Potsdam to Berlin, along with his four sons. As for honorary posts, in 1933 Bieberbach had already been for several years the secretary of the German Mathematical Association DMV, and the editor of the *Jahrbuch über die Fortschritte in der Mathematik*, a review journal which was rivalled since 1930/31 by the newly founded *Zentralblatt*. As of late 1933, he quickly became the leading figure of a group of mathematicians that I will simply call the "NS-mathematicians." Most of them were storm troopers, i.e., members of the SA. Inspired by the concept of national-socialism as a sort of cultural revolution, their goal was to somehow introduce Nazi politics right into mathematical life in Germany. Concretely, this meant that political criteria could override mathematical quality judgements at the institutes controlled by NS-mathematicians – we will see what this meant for the Berlin Institute below. The NS-mathematicians led by Bieberbach tried to gain control of the German Mathematical Association in September 1934, but this attempt had definitely failed by January 1935. After this, Berlin became the central refuge for the more visible NS-mathematicians, in particular from Göttingen: Oswald Teichmüller, Erhard Tornier, and Werner Weber; and also Harald Geppert from Gießen. Only Udo Wegner, a former student of Courant's, managed to run himself a similarly politicized institute, in Heidelberg as of 1937.

The continuity of these political battles with those of the 1920s does not reside in the precise ideological content – at least Bieberbach discovered Nazi slogans for himself only in 1933. But it comes out clearly when one compares the front lines of the thirties to those of the twenties, in particular around the big public controversy concerning the participation of German mathematicians at the 1928 ICM in Bologna. This was the first ICM to which German mathematicians were again invited, i.e., it marked the end of the post–World War I anti-German boycott at this level. Still, the question of whether one should actually follow this invitation generated strong emotions, with the feelings against participating fueled by a national pride which had been frustrated for a long time. Finally, Hilbert led a delegation of 76 mathematicians to Bologna (the biggest national group at this congress), but none of the Berliners went along. The opposition to Hilbert (whose opinion was shared by most mathematicians at Göttingen) was led by the Germanophile Dutch topologist L.E.J. Brouwer and by Bieberbach. Thus three

Theodor Vahlen

basic conflicts merged in this controversy: the question of how much openness to international mathematical relations was adequate for German mathematicians, after years of postwar isolation and ill feelings; the longstanding rivalry between Berlin and Göttingen; and the debate about intuitionism vs. formalism, which Bieberbach could share in, on Brouwer's side, thanks to his predilection for the geometric approach over the algebraic one, and his emphasis of the role of intuition in mathematics. In this perspective, Bieberbach's sudden conversion to Naziism appears as the attempt to replay the old battle, taking advantage of the new distribution of power in Germany.

If such was Bieberbach's plan, he probably misjudged the real balance of power. Ever since the so-called *Röhm-Putsch* (30 June–2 July 1934), i.e., the assasination of 85 SA leaders upon Hitler's command, it became increasingly clear that the new regime had no intention to let Nazi "revolutionaries" do things their way. So without official backing, things were going to be difficult for Bieberbach. Such a backing in mathematics policy would most naturally have to come from the Ministry of Education, or possibly the Academy. For a while it did look like Bieberbach would get all the support he wanted from Theodor Vahlen.

The Greifswald mathematician Theodor Vahlen had been a Nazi activist from as early on as the winter of 1923/24 – see [11], [12]. In 1924, he was *Prorektor* of Greifswald University, and on the anniversary of the (Weimar) constitution that year he ordered the republican flag to be taken down. For once, the Weimar republic reacted strongly, and Vahlen was dismissed from his professorship (without pension rights) in 1927, in spite of numerous solidarity appeals. It was only in 1930 that he found again a position: at the Technical University in Vienna. When the Nazis came to power, he was given back his Greifswald position with great pomp. But he did not stay there:

In 1934, Vahlen became the successor of Richard v. Mises as head of the Berlin Institute for Applied Mathematics. And a few months later he accepted a high-ranking position at the Ministry. Probably upon Vahlen's recommendation, the applied institute was then taken over by the astronomer Alfred Klose who soon reduced it to a negligible quantity.

Once installed as what appeared to many colleagues to be the official Nazi mathematician at the Ministry, Vahlen was thought to back Bieberbach's campaigns. In fact, Bieberbach and Vahlen founded a new journal: *Deutsche Mathematik*, which published (in neatly separated categories) high-level research papers (for instance, many of Teichmüller's works) along with explicitly ideological pamphlets. The journal appeared from 1936 through 1943.

But the Ministry was not reduced to Vahlen, and people there seem to have felt the need for an internationally respectable Mathematical Association, which debates over the Aryan way of introducing π were not promising to create. Thus Bieberbach's influence on the nationwide scale, as well as the funding of *Deutsche Mathematik*, dwindled as time went on, and the Berlin Mathematics Institute became the only true sphere of influence where he could operate freely. It is instructive to see how he acted there.

If one tries to grade all mathematics departments in Hitler's Germany according to how strongly politicized they were, Bieberbach's institute in Berlin and Wegner's in Heidelberg are probably the worst cases. We shall give some concrete examples of what we mean. Before doing so, however, one should point out that such a grading does not mean that Nazi influence at other places was negligible or harmless. It just took a different form, and the variation was quite impressive. In fact, concentrating on Bieberbach too much can divert attention from and even exculpate others by comparison who had not been *like him*, or had even been explicitly against him in the controversy about the DMV.

To take an example, in Berlin there was the assistant K. Molsen who submitted a *Habilitationsschrift* in number theory which Bieberbach and Werner Weber were ready to accept in view of the political qualities of the candidate. It took Erhard Schmidt's intervention to stop this procedure. This certainly shows a failure of responsible action on Bieberbach's part. But on this account he is hardly worse than many others. Helmut Hasse in Göttingen, for instance, tried to propose the *Habilitation* of his assistant Paul Ziegenbein, even though the work was clearly not sufficient. Here, the plan was duly stopped by Carl Ludwig Siegel and Gustav Herglotz.

But there were other things happening in Berlin: not only did Bieberbach propose to the Ministry in 1935 to make Schur an emeritus, but he subsequently asked for Schur's chair to be suppressed completely, giving as a reason that Werner

Weber's lectures ensured adequate teaching in algebra. Given Werner Weber's modest mathematical qualities, this claim is preposterous – unless one reads it as an indirect expression of Bieberbach's contempt for this side of mathematical education.

As mentioned before, Oswald Teichmüller came to Bieberbach's institute for political reasons, thus offering Berlin the presence of the biggest young talent of his generation. It is likely that Teichmüller's switch from algebra (which he had done in Göttingen under Hasse's guidance and in close contact with Ernst Witt) to quasiconformal mappings and then "Teichmüller theory," was also motivated by ideological considerations. But Teichmüller never held a position at the institute; he lived on a modest stipend. This might be explained by Teichmüller's young age, his abrasive personality and extreme working habits, which may have advised against giving him a position involving everyday chores. Still, despite their Nazi affinities, Bieberbach did not succeed in actually integrating Teichmüller into institute life the way he had been involved in seminars at Göttingen. In what almost appears as a conscious suicide, he volunteered in 1943 to the crumbling Eastern front (although they had found a secure job for him in deciphering at Berlin) where he apparently perished sometime in September that year.

Another NS-mathematician from Göttingen, Erhard Tornier, who had previously contributed to foundational problems of mathematical probability, soon developed such an extravagant lifestyle, and ran up corresponding debts, in the capital that he lost both his professorship (it was in fact Landau's chair, which had been transferred for Tornier from Göttingen to Berlin) and his party membership before the war. This gave him the possibility to claim after the war that he had "turned away" from the Nazis early on.

It was also apparently for political reasons that Bieberbach brought to Berlin the Austrian set theorist Foradori as a lecturer. Or at least, if this move had mathematical motives, this would imply a lack of mathematical judgement which is difficult to attribute to Bieberbach.

It also seems characteristic of Bieberbach's handling of things that a number of very talented young mathematicians – like Collatz, Grunsky, Rinow, and Wielandt – were not given regular positions at the institute. Some of them were working for the *Jahrbuch Fortschritte der Mathematik* which Bieberbach continued to edit on behalf of the Berlin Academy.

There is another shameful aspect to the history of the mathematical community in the Third Reich which is immediately connected to Berlin, although not to Bieberbach's institute: the role of the *Reichsverband Mathematik*, which was directed ever since its foundation in 1921 by Georg Hamel from the Technical University Berlin – see [2]. It had been created as a kind of public relations agency for the mathematicians in relation to governments. Consequently, as of 1933, Hamel smoothly switched to Nazi rhetoric, insisting for instance on an alleged *Geistesverbundenheit der Mathematik mit dem Dritten Reich* (spiritual bond between mathematics and the Third Reich) [1], p. 309. But it is not for such rhetorics that the role of Hamel's *Reichsverband* constitutes such a dark chapter in the history of mathematics – a chapter, incidentally, of which research mathematicians at the universities hardly took notice. What the *Reichsverband* did among other things was, in fact, to draw up model exercises for schoolbooks, like these [2]:

> Exercise 85. In the German Reich the population density per 1 km^2 of territory is 140 inhabitants, in Poland 80, South-East-Europe 48, North-East-Europe 32, Soviet Russia 8. – Represent these by five squares of which a part corresponding to the population is shaded.
>
> Exercise 95. The construction of a mental asylum cost 6 million Reichsmark. How many residential houses of 15000 RM each could have been built from this money.

We conclude this chapter by mentioning a few aspects that we have not gone into:

- the decreasing number of mathematics students, and the politization of student life – see the remarks in [6], section 3, and references given there;
- the effect of Nazi politics on mathematical review journals – see [15];
- the role of various government organizations for mathematics and mathematicians in Berlin like the Ministry of Education, the *Deutsche Forschungsgemeinschaft*, etc. – see [12], also [15];
- the war research groups centralized in Berlin – see [3] for a first survey of applied mathematics under Hitler; more historical research on this is desirable.

References

[1] *Hamel, G.*, Die Mathematik im Dritten Reich, UNM **39** (1933), 306–309.

[2] *Mehrtens, H.*, Die "Gleichschaltung" der mathematischen Gesellschaften im nationalsozialistischen Deutschland, Jahrbuch Überblicke Math. (1985), 83–103.

[3] *Mehrtens, H.*, Angewandte Mathematik und Anwendungen der Mathematik im nationalsozialistischen Deutschland, Geschichte und Gesellschaft **12** (1986), Wissenschaften im Nationalsozialismus, 317–347.

[4] *Mehrtens, H.*, Ludwig Bieberbach and "Deutsche Mathematik", in: Phillips, E.R. (ed.), Studies in the History of Mathematics, Studies in Mathematics **26**, Math. Assoc. America (1987), 195–241.

[5] *Merzbach, U.*, Robert Remak and the estimation of units and regulators, in: Demidov, S.S., Folkerts, M., Rowe, D.E., Scriba, C.J. (eds.), AMPHORA, Festschrift für Hans Wußing zu seinem 65. Geburtstag, Birkhäuser, Basel, Boston, Berlin, 1992, 481–522.

[6] *Schappacher, N., (unter Mitwirkung von Martin Kneser)*, Fachverband – Institut – Staat, Streiflichter auf das Verhältnis von Mathematik zu Gesellschaft und Politik in Deutschland seit 1890 unter besonderer Berücksichtigung der Zeit des Nationalsozialismus, in: Fischer, G., Hirzebruch, F., Scharlau, W., Törnig, W. (eds.), Ein Jahrhundert Mathematik 1890–1990, Festschrift zum Jubiläum der DMV, Vieweg, Braunschweig, 1990, 1–82.

[7] *Schappacher, N., Scholz, E. (eds.)*, Oswald Teichmüller – Leben und Werk, mit Beiträgen von K. Hauser, F. Herrlich, M. Kneser, H. Opolka, N. Schappacher, E. Scholz, Jber. dt. Math.-Verein. **94** (1992), 1–39.

[8] *Siegmund-Schultze, R.*, Ein Mathematiker als Präsident der Berliner Akademie der Wissenschaften in ihrer dunkelsten Zeit, Mitt. Math. Ges. DDR **2** (1983), 49–54.

[9] *Siegmund-Schultze, R.*, Das Ende des Jahrbuchs über die Fortschritte der Mathematik und die Brechung des deutschen Referatemonopols, Mitt. Math. Ges. DDR **1** (1984), 91–101.

[10] *Siegmund-Schultze, R.*, Einige Probleme der Geschichtsschreibung der Mathematik im faschistischen Deutschland – unter besonderer Berücksichtigung des Lebenslaufes des Greifswalder Mathematikers Theodor Vahlen, Wiss. Z., Ernst-Moritz-Arndt Univ. Greifswald, Math.-Nat. Reihe XXXIII (1984), 51–56.

[11] *Siegmund-Schultze, R.*, Theodor Vahlen – zum Schuldanteil eines deutschen Mathematikers am faschistischen Mißbrauch der Wissenschaft, NTM-Schriftenreihe für Geschichte der Naturwissenschaften, Technik und Medizin **21** (1) (1984), 17–32.

[12] *Siegmund-Schultze, R.*, Faschistische Pläne zur "Neuordnung" der europäischen Wissenschaft, Das Beispiel Mathematik, NTM-Schriftenreihe für Geschichte der Naturwissenschaften, Technik und Medizin **23** (2) (1986), 1–17.

[13] *Siegmund-Schultze, R.*, Berliner Mathematik zur Zeit des Faschismus, Mitt. Math. Ges. DDR **4** (1987), 61–84.

[14] *Siegmund-Schultze, R.*, Zur Sozialgeschichte der Mathematik an der Berliner Universität im Faschismus, NTM-Schriftenreihe für Geschichte der Naturwissenschaften, Technik und Medizin **26** (1) (1989), 49–68.

[15] *Siegmund-Schultze, R.*, Mathematische Berichterstattung in Hitlerdeutschland – Der Niedergang des "Jahrbuchs über die Fortschritte der Mathematik", Studien zur Wissenschafts-, Sozial- und Bildungsgeschichte der Mathematik 9, Vandenhoeck & Ruprecht, Göttingen, 1993.

The University of Berlin from Reopening until 1953

Reinhard Siegmund-Schultze

When the University of Berlin – only in 1949 to be named after the Humboldt brothers – resumed its activities on January 29, 1946, not one of its renowned mathematicians of the 1920s and 1930s was still in office. Many prominent mathematicians, among them the full professors Richard von Mises and Issai Schur, had been expelled by the Nazis; the latter had died meanwhile in Palestine. The most outspoken supporters of Hitler's regime among the mathematicians had been dismissed and arrested (L. Bieberbach), had died in the turmoils immediately after the war (Th. Vahlen), had committed suicide (H. Geppert), or had been killed as war volunteers (O. Teichmüller). The applied mathematician A. Klose had been sent to the Soviet Union for reparation work. The co-founder of functional analysis, Erhard Schmidt, who had been politically moderate during the Nazi years without openly resisting the regime, did not return to the university until late 1946. The algebraist Hermann Ludwig Schmid – by his own testimony "the only German mathematician in Berlin who gained a docentship between 1933 and 1945 without joining the Nazi party" – tried to resume teaching as early as May 1945. Initially he operated from a Western sector of Berlin, soon to be occupied by the Americans, where the bigger part of the mathematical institute had been moved due to the war-related demolitions in the East. The relocated mathematical library came back to Berlin in late 1945. Running the institute from West Berlin created additional problems given the growing political tension between the Western and Eastern parts of Berlin.

It took eight years before the mathematical institute would resettle in full in the restored main building of the university in East Berlin in 1953. Besides Schmid the mathematical educator F. Neiss and the Greek-born function theorist Alexander Dinghas were granted teaching permits in 1946. Due to his membership in the *Nationalsozialistische Deutsche Arbeiterpartei* (NSDAP) the teaching permit of the aerodynamicist Kurt Schröder was temporarily revoked. Thus, it became one of the most pressing needs of the re-opened university to find politically relatively untainted professors. The managing director of the university was sent on a trip to the Western zones of Germany in December 1946 in order to "negotiate the conditions under which several gentlemen, who have been called to the University or to the Academy of Sciences, are willing to come to Berlin." The negotiations with C. Carathéodory, G. Hamel, M. Geppert (in fact a woman) and M. Deuring came to no result. The only appointment which could be made was that of the renowned number theorist Helmut Hasse – but only because he had been dismissed in the Western zones due to his, at least, temporary sympathy with the Nazi regime. The appointment of the actuarial mathematician, P. Lorenz, was supported in 1949 in

Opening ceremony of the university in 1946

spite of political and scientific reservations only on the grounds that "there is no equally qualified person (as M. Geppert) available."

In 1948, Emmy Noether's student Heinrich Grell came from Erlangen in West Germany. Schmid recommended him as "the one German mathematician who suffered the strongest persecutions by the past regime," thereby obviously ignoring the much more severe sufferings of the emigrants and other victims of the Nazis. In 1952, number theorist Hans Reichardt, who had served as a "specialist" in the Soviet Union, joined the staff of the mathematical institute, where he soon developed differential geometry as his second field of interest. The mathematicians as well as professors in other fields received high salaries and extra benefits in order to prevent them from leaving East Germany. From the outset the development of mathematics in Berlin, both at the university and the Academy of Sciences, was strongly supported by the Soviet occupation authorities and (after 1949) by the new government of the German Democratic Republic. While the mathematical institute at the university aimed initially (in 1946) at "educating teachers for the upper grades of the German Standard School (*Deutsche Einheitsschule*)," that goal was supplemented in January 1948 with regard to "educating students for jobs that need mathematical university training." In April 1949 the German Administration for Education in the Soviet Occupation Zone (*Deutsche Zentralverwaltung für Volksbildung in der Sowjetischen Besatzungszone*) accepted a new curriculum for "Practical Mathematics." Twenty-one students of mathematics were admitted in the winter term 1949/50, more than in physics and much more than in biology. Finally, the mathematical research profile was enriched by the foundation in January 1951 of an "Institute for Mathematical Logic" under Karl Schröter.

Hans Reichardt, Helmut Hasse

The political positions of mathematicians and science functionaries were partly in agreement, partly in conflict. Erhard Schmidt, whose 75th birthday on January 13, 1951, was celebrated by a large assembly of guests from all over Germany, saw a Pan-German responsibility for mathematics at the traditional University of Berlin. This responsibility was underscored by the political authorities in their strategies with respect to the Academy of Sciences. Schmidt, who received the new National Prize of the GDR in 1949, disapproved of Dinghas' departure for the *Freie Universität* in the American sector of Berlin, which had been founded in 1948. Schmidt considered the German Mathematicians' Association (*Deutsche Mathematiker-Vereinigung*), re-founded by the mathematician E. Kamke from Tübingen in West Germany, as one of the "braces which hold together a common German culture." Kamke, who had been a victim of the Nazis, condemned the attempts of West Berlin mathematicians, such as Dinghas, to rehabilitate Bieberbach, the politically most incriminated of all German mathematicians. For that reason, and due to the permanent political confrontation between the Western and Eastern parts of Berlin it was not surprising that the relations between the mathematicians at Humboldt University and their West German colleagues would be in some respects closer and more cordial than the relations between East and West Berlin mathematicians. On the one hand this accounted for some political agreement between the mathematicians and the communist authorities. On the other hand, however, there were early conflicts between the two parties. Conflicts arose partly from the very conservative positions of mathematicians such as Erhard Schmidt, who opposed in 1947 a lectureship for the statistician E.J. Gumbel, which had been proposed by the administration. Gumbel had been one of the few leftist scholars and outspoken defenders of the Weimar Republic; in the 1950s he would have a temporary appointment at the *Freie Universität*. An even more important source of conflicts with the mathematicians were the admission and curricula policies of the communist science functionaries.

The preferential admission of children from disadvantaged social strata met with some approval even from West German function theorist H. Behnke, who was on a short visit to East Berlin in November 1947. However, it frequently resulted in barring gifted applicants. This circumstance led Hasse in 1949 to demand an "exemption clause for the extraordinarily gifted". Mandatory lectures for students on "Political and Social Problems of Today" were introduced as early as 1946. These relatively undogmatic lectures were replaced by much more politicized ones in 1951. Mathematics students had to pass exams on topics such as "Walter Ulbricht's Textbook on the Structure of State and Economy" (Ulbricht being the head of the communist Socialist Unity Party (*Sozialistische Einheitspartei*)).

In 1952, one year before the deep political crisis in the GDR, conflicts between mathematicians and science functionaries intensified. As dean of Rostock University, the former Berlin mathematician R. Kochendörffer protested against the new 10-month curriculum, the exaggerated emphasis on non-mathematical issues in it, and the gradual curtailment of academic self-determination of the universities. Kochendörffer found support among the Berlin mathematicians who had to face additional problems. Grell opposed the arbitrary dismissal of students from West Berlin and declared that, "beyond any political considerations," there were "human points of view which I am not prepared to sacrifice." A permanent source of tension between mathematicians and political functionaries was the shortage of foreign currency in the GDR, which severely limited the purchase of Western literature and the participation in scientific meetings and congresses such as the International Congress of Mathematicians in Cambridge, USA, in 1950. By and large, even in 1953, when the most active administrator among the mathematicians at Humboldt University, H.L. Schmid, left for Würzburg in West Germany, Behnke's words after his short trip to East Berlin in November 1947 were still valid: "I am envious of the much greater official support that the colleagues in Berlin find in normalizing their lives and in the reconstruction of the university. Still, I did not feel comfortable in Berlin. Given the insecure political situation and the unstable economic circumstances, every work effort has a touch of eeriness. The increasing strangulation of bourgeois circles and, what is more, of bourgeois culture, in the Russian zone makes intellectual life very volatile. The Germans in the West will have the difficult choice between a continued embracement of the European culture and the unity of Germany."

Unpublished sources

Papers of Kurt Schröder and Karl Schröter in the archives of Berlin-Brandenburg Academy of Sciences; files from the archives of Humboldt University, especially the Faculty of Sciences, and personal files of K. Schröder, K. Schröter, H.L. Schmid, E. Schmidt, P. Lorenz, A. Dinghas, H. Hasse, H. Grell, R. Kochendörffer, H. Reichardt; report by H. Behnke, November 1947 (5 p.) from: Oswald Veblen Papers, Library of Congress, Washington, D.C., file: R. Courant.

References

[1] *Siegmund-Schultze, R.*, Dealing with the political past of East German mathematics, The Mathematical Intelligencer **15** (4) (1993), 27–36.

[2] *Siegmund-Schultze, R.*, Zu den ost-westdeutschen mathematischen Beziehungen bis zur Gründung der Mathematischen Gesellschaft der DDR 1962, Hochschule Ost **5** (1996), Heft 3, 55–63.

[3] *Siegmund-Schultze, R.*, The shadow of National Socialism: Political and Psychological After-Effects of Hitler's Regime on Science in the GDR, in: Macrakis, K., Hoffmann, D. (eds.), Science in the GDR, Cambridge, Mass., Harvard University Press, 1998 (to appear).

Helmut Hasse, Hermann Ludwig Schmid and their students in Berlin

Wolfram Jehne and Erich Lamprecht

In the early post-war years two personalities had substantial influence on the scientific and institutional development of mathematics in Berlin: Helmut Hasse (1898–1979) and Hermann Ludwig Schmid (1908–1956). In the following the authors, former students of Hasse and Schmid, respectively, try to create from a personal point of view a picture of the achievements reached in those most difficult times. For a correct understanding of the general and academic situation in post-war Berlin it seems appropriate to recall the political status of the city.

After the war Berlin was divided into four sectors, each of them controlled by one of the former allies. Due to the tensions between Western Allies and the Soviet Union immediately following the war, the political situation steadily deteriorated. In 1948 this development culminated in the first of a series of events which marked the division of the city into East Berlin and West Berlin, namely the introduction of two different currencies: In June, West Berlin joined the currency reform of West Germany (*West-Zone*). A few days later the Soviet administration introduced a separate currency in East Germany (*Ost-Zone*) and East Berlin and imposed the blockade on West Berlin. About the same time, a freely elected common city assembly with residence in East Berlin was forced to end its activities there; the Western delegates had to move to West Berlin. Thus, the division of the city had been completed also on the administrative level. Earlier that year, the Soviet administration had already left the Control Council.

It was towards the end of this year that the *Freie Universität Berlin* (Free University of Berlin) was founded in West Berlin and the traditional former *Friedrich-Wilhelms-Universität* in East Berlin obtained its present name, *Humboldt-Universität*. One year later the two German states were established, the Federal Republic of Germany and the German Democratic Republic (BRD and DDR). These had been the general political settings in which the rebuilding of the widely destroyed city and the reorganisation of the economical and cultural infrastructure took place. Academic structures and academic life developed rather quickly whereas private life took longer to normalize.

The former *Friedrich-Wilhelms-Universität* reopened in January 1946, for the time being under the neutral name *Universität Berlin*. The Mathematics Department had rooms in the west wing of the damaged main building *Unter den Linden*, East Berlin, and a branch office in West Berlin.

H.L. Schmid was in Berlin at that time and he became quickly one of the main figures in the Department, combining high determination with diplomatic skills. Having been a student of Hasse, he became his assistant in Göttingen between 1935 and 1937, after he had written his thesis with him in Marburg in 1934. In

Helmut Hasse

1938 he became editorial secretary of the *Zentralblatt für Mathematik* in Giessen, where he also habilitated. Schmid finally moved to Berlin in 1940, starting as an assistant of H. Geppert until he became *Privatdozent* (lecturer).

Schmid was appointed full professor in 1946 and soon became a member of the *Gelehrter Rat für Fragen der wissenschaftlichen Forschung und Lehre*, an official consulting council at the *Zentralverwaltung für Volksbildung* (see [3]). Thus, he gained the influence needed to realize his plans, to the benefit of the Department and of mathematics in Berlin.

At an early stage Schmid organized the beginning of teaching. The first official beginner courses started in February 1946, whereas the Department had taken up its work in the branch office unofficially as early as May 1945, immediately after the war.

It seems remarkable that it was possible in those times to provide several visiting professorships for full-term courses at the *Humboldt-Universität*. The geometer Eduard Rembs and the mathematical physicist Georg Hamel from West Berlin, the analyst Konrad Knopp, and the algebraist Friedrich Karl Schmidt from West Germany accepted invitations, the latter even for a whole year in 1947/48. F.K. Schmidt's one-term lectures on mathematical logic indeed were highlights in those days. His way of celebrating the subject attracted many students also from other faculties, not only mathematicians and physicists. Later, he obtained an offer from the *Humboldt-Universität* which he rejected.

Hermann Ludwig Schmid

Thanks to the efforts of H.L. Schmid the prominent mathematicians Erhard Schmidt, former rector of the university, and Helmut Hasse, both of whom lived in West Germany, could be won to join the Department in 1946. For an extensive appreciation of Erhard Schmidt we refer to the article of Baumgärtel in this book. The call to Berlin suited Hasse well as he had been dismissed in Göttingen by the British authorities.

In order to characterize Hasse as a mathematician one may quote the first sentence from his short biography by H.M. Edwards (cf. [1]): "One of the most important mathematicians of the 20th century, Helmut Hasse, was a man whose accomplishments spanned research, mathematical exposition, teaching, and editorial work."

In the twenties and thirties, Artin and Hasse had been the leading representatives of algebraic number theory in Germany, though of course, the work of Richard Brauer and Emmy Noether had great influence on the discipline as well. We mention just some of Hasse's great works of that period: His proof of the Riemann hypothesis for congruence function fields in the elliptic case, and the *Local-Global Principles* – later named after him – in different branches of number theory, namely for quadratic forms, for the arithmetic of algebras, and for norm-maps. Much later, Hasse principles and obstructions to them were to turn up in various contexts of modern mathematics. Some of Hasse's Berlin students also contributed to phenomena of that type (M. Kneser, P. Roquette, W. Jehne).

In 1946, the *Deutsche Akademie der Wissenschaften zu Berlin* (German Academy of Sciences in Berlin) was founded. On the initiative of H.L. Schmid, a Research Institute for Mathematics was opened as an Institute of the Academy with E. Schmidt as director. In 1947, Schmid succeeded in reinstalling the German Review Series *Zentralblatt* and took over the editorial responsibilities. About the same time Schmid also initiated the resumption of the project *Mathematisches Wörterbuch* (Dictionary of Mathematics). When in 1947 the periodical *Mathematische Nachrichten* was founded, it was again H.L. Schmid who had initiated and promoted the project. Erhard Schmidt became chief editor, whereas Schmid acted as managing editor.

H.L. Schmid's mathematical research was devoted to various topics in algebraic number theory and classical algebraic geometry. His early work concentrated on function fields and Gauss sums. During his years in Göttingen (1935–37) he belonged to the seminar of Hasse on congruence function fields and p-adic numbers, together with E. Witt and O. Teichmüller. A close collaboration between Schmid and Witt led the latter to what is now known as Witt vector calculus. Later, in Berlin, Schmid published several papers on differential equations of Lamé type and on algebraic questions related to them.

The way Schmid collaborated with his research students Arno Cronheim, Erich Lamprecht, Heinz Orsinger (assistant), and his collaborator Otto Grün, and later on Helmut Boseck, was based more on team work and private discussions than on special lectures and research seminars. The research projects naturally stemmed from Schmid's own field of research: classical algebraic geometry, mainly curves, and differential equations. Lamprecht started with an extensive treatment of Gauss sums in finite rings (thesis 1952). Later on he turned to algebra and to algebraic function fields and studied their reduction and the corresponding zeta functions.

When Hasse decided to go to Berlin he was nearly 50 years old. During his Berlin years he lived in West Berlin. To live in West Berlin and work or study in East Berlin was still possible in those years, and many professors and students of the *Humboldt-Universität* did so. Hasse first held a research position at the Academy before starting to teach in the summer term 1948 with a three-term course on elementary and algebraic number theory, followed by a one-term course on arithmetics of algebras and a two-term course on complex multiplication. His courses were accompanied by seminars, for instance on diophantine approximations and class field theory. In two and a half years of teaching Hasse guided a group of research students to the new developments and open problems in algebraic number theory.

It was a large group of students who had been attracted by Hasse's personality and his mathematics: Herbert Benz (†1988), Gudrun Beyer, Wolfram Jehne, Hans-Wilhelm Knobloch, Heinrich-Wolfgang Leopoldt, Paul Wolf, Peter Roquette and

Martin Kneser. The two latter students came from West Germany in the critical years 1948/49 in order to work with Hasse. Curt Meyer, personal assistant of Hasse, was responsible for problem sessions, seminars and close contact to the students.

The lectures and regular seminars were held in the main building *Unter den Linden*, whilst the research seminars took place in Erhard Schmidt's private apartment in West Berlin. They focused on recent results, in particular unpublished papers of Hasse were discussed.

Hasse published three books during his stay in Berlin: a handbook entitled *Zahlentheorie* (1949), his *Vorlesungen über Zahlentheorie* (1950) which had emerged from the course mentioned above, and a monograph with the title *Über die Klassenzahl abelscher Zahlkörper* (1952). It was this last book which inspired H.-W. Leopoldt to start his deep investigations on abelian fields which finally led him to p-adic L-functions and the Galois module structure of abelian integers. Leopoldt thus gave decisive impetus to the development of Iwasawa theory and Galois module structure theory. He actually was one of their co-founders.

C. Meyer's contributions to abelian extensions of quadratic fields with emphasis on explicit class number formulae (monograph 1957) were also influenced by Hasse. In the imaginary case complex multiplication was at hand; the much more intricate real case relied on analytic techniques introduced by Herglotz and Hecke.

M. Kneser drew up Hasse's course on complex multiplication, but a printed version never appeared. When Shafarevich's result on a general reciprocity law appeared Kneser promptly published a simplified version and added the missing case for the prime 2 (1951). Nevertheless, it so happened that he wrote a geometric dissertation under Erhard Schmidt.

Hasse was deeply pleased when P. Roquette, in 1951, gave a proof of the Riemann hypothesis for congruence zeta functions in the higher genus case in the framework of Hasse's and Deuring's previous work. A more geometrically inspired proof was given by A. Weil in 1948. In his extensive later work Roquette also had looked at Hasse principles from the viewpoint of model theory, particularly decidability questions. He thereby shed a new light on Hasse's early achievements.

Other students of his made their first mathematical steps in connection with Hasse's actual research in Berlin: embedding problems for fields and nonabelian Kummer theory (see [4], vol. 2). The latter theory was generalized and simplified by P. Wolf (monograph 1956). An arithmetical refinement was given by H.-W. Knobloch, who proved some kind of nonabelian decomposition law (thesis 1950). After further work on analytic number theory, Knobloch changed subjects completely and turned to dynamical systems and control theory. The work on embedding problems had two aspects: the characterization part and the existence part. Hasse's general conjecture concerning existence was disproved by Faddeev. But G. Beyer could show a modified conjecture to be true for a large class of special situations (thesis 1956). As to the characterization part, W. Jehne solved Hasse's *Widerspiegelungsproblem* by embedding global algebras into adèle algebras (1952). The result is nowadays referred to as the theorem of Shafarevich-Weil.

I.R. Shafarevich, B.N. Delone, S.P. Demushkin, A.V. Jakovlev and D.K. Faddeev belonged to an important Russian school of mathematicians who also worked on embedding problems and related questions, at about the same time. There can be no doubt that in particular Shafarevich's striking results were outstanding achievements in algebraic number theory. However, the two protagonists, Hasse and Shafarevich, came to meet only as late as 1959 in Berlin. And it was in 1964 that Shafarevich was able to attend the annual meeting on algebraic number theory in Oberwolfach, for the first and only time.

In a long programmatic preface to his book on class numbers Hasse presented his viewpoint on constructiveness in arithmetic, very much in the spirit of classical authors like Gauss, Kummer and Kronecker (see [5]). It has influenced algebraic number theory right up to the development of modern *Constructive Number Theory* (Leopoldt, Zassenhaus et al.). Among the numerous papers Hasse wrote in Berlin one also finds some in this spirit, e.g., contributions to the arithmetic of small fields. Hasse's monograph on Galois Gauss sums (root numbers) was applied much later in Fröhlich's Galois module structure theory. And his contribution to the so-called *Hasse-Weil zeta functions* of global Fermat-type function fields complemented Weil's important papers on the same subject. These examples may illustrate the widespread range of Hasse's work in Berlin (see [4]).

The group theorist Otto Grün (1888–1974), an interesting figure and outsider in Berlin, also deserves attention. Though he cannot be considered a student of Hasse or Schmid his fate as a mathematician was tied to both. He originally had been a bank clerk, in mathematics he was entirely self-educated. Hasse discovered him when he accepted his paper on the Fermat problem for publication in *Crelle's Journal* in 1932. According to Grün, Hasse had advised him to switch to group theory. In five papers entitled *Beiträge zur Gruppentheorie* (Contributions to Group Theory) he proved his celebrated theorems which nowadays belong to the standard results in elementary group theory. He had no permanent teaching obligations though attempts had been made. As of 1947, his source of living was an Academy research grant which H.L. Schmid could provide in Berlin and later also in Würzburg.

In 1950 Hasse accepted a chair at Hamburg University and encouraged his students to follow him. Since he was able to raise small grants and to provide accommodation in Hamburg, everybody, even the East Germans among his students, could grasp the opportunity. For the rest of his life the small town Ahrensburg near Hamburg remained Hasse's place of residence. From there he maintained contact with his colleagues and friends of the Berlin Academy for a long time.

Hasse was a man of great personal charisma, and he was a fascinating and generous teacher. Up to his death in 1979 he had a deep concern for music and used to play Bach and Beethoven, in particular, in his Berlin and early Hamburg years Beethoven's late piano sonatas.

In 1953 H.L. Schmid was offered a chair at Würzburg University which he accepted. Preceeding negotiations with the Bavarian administration had been concluded to his satisfaction, and many of his collaborators could follow him from

Berlin to Würzburg, Lamprecht and Klaus Krickeberg – a student of E. Schmidt – among them. Schmid immediately concentrated his efforts on the planning of the construction of a new building for the Mathematics Department and on establishing a research section with numerous non-resident collaborators who continued his extensive mathematical work; see *Mathematische Nachrichten* 18/19 (1958). Soon, Schmid became Dean of the Faculty of Science and finally Rector of the University. In 1956 he died from a sudden heart attack.

The merits of Hasse and H.L. Schmid during their time in Berlin may be highlighted in the following way: Schmid brought mathematical life into existence by creating the institutional framework in those most difficult post-war years. Hasse's four years in Berlin were very fruitful ones: he attracted a group of devoted students, wrote about 25 papers and published three books. Though Hasse's most important work was done in the twenties and thirties, one may say that his research in Berlin came to another culminating point.

Altogether, in cooperation with Erhard Schmidt and the other colleagues, Hasse and H.L. Schmid, each in his very personal way, led mathematics in Berlin to a first revival.

References

[1] *Edwards, H.M.*, Helmut Hasse, Dictionary of Scientific Biographies.

[2] *Frei, G.*, Helmut Hasse (1898–1979), Expositiones Mathematicae **3** (1985), 55–69.

[3] *Hasse, H.*, Wissenschaftlicher Nachruf auf Hermann Ludwig Schmid, Math. Nachr. **18** (1958), 1–18.

[4] *Hasse, H.*, Mathematische Abhandlungen, 3 vols., de Gruyter Verlag, Berlin-New York, 1975.

[5] *Leopoldt, H.-W.*, Zum wissenschaftlichen Werk von Helmut Hasse, J. Reine Angew. Math. **262** (1973), 1–17.

[6] *Leopoldt, H.-W.*, Obituary (for Helmut Hasse), J. Number Theory **14** (1982), 118–120.

[7] *Roquette, P.*, In Memoriam Helmut Hasse, Manuscript 1980, 1–4.

[8] *Schmidt, H.*, Hermann Ludwig Schmid, Jber. dt. Math.-Verein. **61** (1958), 7–11.

Freie Universität Berlin
A summary of its history

Heinrich Begehr

Among all German universities there are very few whose history is closely related to the Germany history after the Second World War. One is the *Humboldt-Universität zu Berlin*; the other is the *Freie Universität* (Free University) of Berlin which is unique for another reason: It is the only university in the whole world that was founded by a student initiative.

Origin

In 1948 the Free University separated from the *Berlin Linden Universität* as the well-known *Friedrich-Wilhelms-Universität* was briefly called after the war before it was renamed *Humboldt-Universität* in 1949. Some remarks have to be made about this institution.

Already the foundation of the *Friedrich-Wilhelms-Universität zu Berlin* in 1810 was politically motivated. France under Napoléon had conquered Prussia, occupied the country and closed the universities to the west of the Elbe River. The newly founded University of Berlin was supposed to facilitate a spiritual and moral renewal for the depressed Prussian population. And it was to foster spiritual independence of the citizens of Berlin. Acting in combination with the Prussian Academy of Sciences research and teaching should be carried out. Wilhelm von Humboldt was the spiritual father and Alexander von Humboldt started the new age of natural sciences in Berlin in 1827 with his lectures about the cosmos.

In the second part of the last century the University of Berlin became the leading university not only in Germany. For many fields, including mathematics, it became the center of the scientific world. In connection with the *Kaiser-Wilhelm-Gesellschaft* founded in 1910 and nowadays called the *Max-Planck-Gesellschaft*, Berlin kept its extraordinary importance in sciences. In the twenties nowhere else in the world could one find more Nobel prize winners than in Berlin.

National socialism caused a sudden end to this flourishing scientific atmosphere. Einstein's leaving Berlin in 1933 was a striking sign of the changing situation. After the war, the University of Berlin was reopened in February 1946, then cleansed of former national socialists. The end of the Second World War and the breakdown of the political regime had triggered great expectations, especially in the young generation. Many of the new students had served as soldiers or had joined resistance groups during the war. They had faced political pressure and had missed freedom of speech. Hence, at the end of the war students took to the new spiritual freedom with enthusiasm. But soon in Berlin they met new bounds set by the Soviet occupying power.

When the University of Berlin started to function again officially in February 1946, it was not only reorganized. Located in the Soviet sector of Berlin, the

Mathematical Institute of FU at Hüttenweg

Soviet military administration had placed the university under their authority. The Western allied forces, the Americans and the British and then also the French, were not well-prepared to take over administration and responsibility in Berlin. Only the Soviets had definite and detailed plans from the beginning. Thus they had taken over the university under their rule before the other allied forces had even given it a thought. In the period which followed, the Soviets not only started to impose political pressure on students who exercised their new freedom and did not collaborate with them. They also founded a new faculty for workers and farmers where students were accepted without high school diploma on the strength of their parents being workers or farmers. These students, of course, were grateful and loyal to the new communist administration appointed by the Soviets. Classes in Marxism-Leninism became obligatory and after a short time only politically loyal students were admitted to register. Finally, students who were disliked by the Soviet administration, especially some Social Democrats and Christian Democrats, were arrested. This began in 1946 and continued in 1947 and 1948. Some of these students just disappeared; others were sentenced to 25 years in hard labour camps by Soviet military tribunals. They were released after about 8 years. Several of them had belonged to the resistance movement against the Nazi regime and had already been in prison under that government.

One tragic example was Georg Wrazidlo. During the war he had received his high school diploma as a sixth-former and was drafted for labour and military services. In 1942 he was admitted to study medicine at the Breslau University. At the same time he was promoted to first lieutenant. In Breslau he joined an antifascist student group formed by some of his former classmates. On these grounds he was arrested in 1944 by the *Gestapo*, the German secret police. After having spent some time in the Buchenwald concentration camp he was placed in a particular company in order to prove his worth. At the end of the war he was shortly under American arrest before he entered the newly opened University of Berlin as one of the first students. He became chairman of the student association. But already in 1946 he was fired from this chair because he had joined the protests of students against the misuse of the university for political party affairs. He was one of the founders of the university section of the CDU, the *Christlich-Demokratische Union* (Christian Democratic Union), and the student speaker at the opening ceremony of the university in February 1946. His later fate, in short: March 13, 1947, arrested in East Berlin; December 13, 1948, given a 25-year prison sentence by a Soviet military tribunal because of spying; forced labour camps in Bautzen and Brandenburg; October 13, 1956, released from prison; final examination in medicine at the *Freie Universität*; 1959, decorated with the order of the Federal Republic of Germany; met his death in a traffic accident.

Foundation

The incident which ultimately incited the students to demand a new free university in the Western zones of the city was the exclusion of three students from the university. They were the editors of a student journal *Colloquium* founded in March 1947, and much disliked by the Soviet administration. Their names were Otto Heß, Joachim Schwarz and Otto Stolz.

April 16, 1948 These students are expelled from the university by its rector.

April 23, 1948 During the subsequent protest demonstration at the Hotel Esplanade in the British sector of Berlin next to the border to the Soviet sector, students effectively propagate for the first time in public their demand for a new – a free – university in the Western part of the city. They want a university free from political pressure, controls and other restrictions. They immediately draw a lot of public attention. The population in West Berlin and the American military administration react positively. At that time the British think about extending the Technical University of Berlin. In Dahlem – a district of the American sector – a German university for research starts with the institutes of the former *Kaiser-Wilhelm-Gesellschaft* following the American model. This university, however, is quickly abandoned because soon the *Freie Universität* and the *Max-Planck-Gesellschaft* settle in Dahlem.

April 24, 1948 Heß, Schwarz and Stolz convince an American journalist of their ideas, who starts to persuade General Lucius Clay, the Chief Commander of the US forces in Berlin. On the following weekend the students start to contact Ernst Reuter, the elected mayor of Berlin, who is not accepted by the Soviets. The growing popularity of the students' demand and the political climate at the start of the Cold War supported their plan.

May 11, 1948 The Berlin city parliament votes in favour with 83:17:5 for a motion by the SPD, the *Sozialdemokratische Partei Deutschlands* (Social Democratic Party of Germany), to found a free university.

June 19, 1948 A group of 40 representatives of public and academic life in Berlin, among them two students and one PhD student, decide to establish the *Freie Universität*. They form a 12-member preparatory committee. Ernst Reuter and many other politicians are unable to participate at this meeting because of the forthcoming currency reform.

September 2, 1948 The Berlin government decides to found the *Freie Universität.*

There were Americans who thought that, under ideal conditions, in the United States it would take two years before a university could begin to operate. In Berlin the conditions were far from ideal. In addition to all the organisational problems, the Berlin blockade was errected by the Soviets on *January 24, 1948.* The new university had three main hurdles to overcome:

- Recruiting of the academic staff – General Clay had demanded a list of 50 professors willing to leave the *Humboldt-Universität* for the new one. In the end, Professor Edwin Redslob, later the first acting Rector of the Free University, was able to present a list of only 30 scientists.
- A legal basis for the new university had to be created which would prevent Soviet interference.
- Financial backing had to be secured.

Of great help was the energetic support from Ernst Reuter, the mayor of Berlin, and from other politicians as well as the positive and detailed reports in the Western newspapers resulting in a positive attitude from the population. By October 1948 about 5000 students had applied for registration.

November 10, 1948 The first teaching period started. 30 % of the first-year students came from East Germany, the others from Berlin. Among the advanced students 70 % came from the *Humboldt-Universität* in East Berlin and 9 % from universities in West Germany. The selection of students from the applications was made in close cooperation with founding students.

The majority of the professors of the *Humboldt-Universität* rejected the founding of the Free University. Most of them had joined the famous *Friedrich-Wilhelms-Universität* years ago, before the Second World War, and they had developed a great loyalty towards it. They thought that founding the Free University was a betrayal of their old one. They felt that in such a difficult time a professor was not supposed to leave his university and his students. They presumed that the more politically independent faculty members would leave, the easier it would be for the new system to force the university in line. Only very few and mostly younger professors went over to the new university. It was assumed that they left because of better living conditions and for a better career.

Such resistance was only one reason for the difficulties in filling the positions. After the war several vacant professorships had to be filled at German universities and the number of candidates was small. Many scientists had been killed, some had emigrated, others could not be re-employed because of their former involvement in the Nazi regime. Some scientists were brought to the Soviet Union, others were hired by the United States, for instance in order to continue the development of rockets. Moreover, the uncertain political situation did not attract people to move to Berlin.

Continuous support from the United States – morally as well as financially – was of great importance. Without it this enormously difficult untertaking would have failed. At the beginning the Americans donated DM 2 000 000. On June 16, 1951, the Ford Foundation helped with US $ 1 309 500 to build the Henry Ford Building, the main building of the Free University, together with the main library and the canteen. Again in 1961 the United States supported the Free University, this time with sixty million US dollars for the Free University Hospital in Steglitz which was later renamed the Benjamin Franklin Clinic.

Within three months, by December 1948, the library of the Free University had collected a stock of about 350 000 books. This made that library bigger than those of many other established universities, whose holdings had been reduced because of the war.

December 4, 1948 The opening ceremony of the Free University was held in the Titania Palace in Steglitz.

Consolidation

The foundation of the Free University on the initiative of students' activities had created a particular model of the Free University administration under essential collaboration with the students. But this model failed to be accepted by other universities in Germany. In the following years this system got more and more disbanded and the Free University was brought into line with the other universities in West Germany. Later, in the sixties, the lack of student participation was one of the major reasons for criticism.

The Free University succeeded in earning scientific profile and appreciation. Its speciality was

- inclusion of the College for Politics as the Otto Suhr Institute in 1958;
- promotion of political and social sciences, which helped German universities to redevelop their capability in these disciplines and made the Free University the leader in Central Europe;
- the Institute for East European and Slavistic Sciences;
- the John-F.-Kennedy Institute for North American Studies where many scientists from America filled positions.

The number of students soon grew. After one year it had as many students as there were at the *Humboldt-Universität* in East Berlin. In 1955 there were about 5000 of them, and many of them came from East Germany. Until 1961 more students came from West Germany while the number of students from East Germany grew only slightly. In the winter term 1960/61 there were 6 % from East Berlin, 30 % from East Germany. The erection of the wall in 1961 caused a big change. Nevertheless the number of students kept growing. In 1962 there were 14 000, in 1976 about 40 000 and in the eighties almost 60 000 students at the Free University. Together with the universities in Munich and Münster, the Free University is one of the biggest in Germany. The Federal Republic of Germany had altogether 290 000 students in 1960 and 400 000 in 1966.

There were no contacts at all to universities in East Germany or in the whole of Eastern Europe. Only in 1968 did first contacts with the Soviet Union become possible and a partnership with the University of Leningrad was established.

Student revolution and reform of universities

With the change of generations the spirit of the founding years disappeared. The Free University was improving academically. Especially political sciences and sociology became important branches; physics and German studies built up good reputations. The building of the Berlin wall created once again anti-communist tendencies among the students.

In 1958 the first student demonstrations arose in connection with the rearmament of West Germany. In the mid-sixties the new youth revolt started mainly at universities in West Germany. Suspicious of any authority, students attacked the West German foreign policy because of its orientation to the West and its non-recognition of the German Democratic Republic. Moreover, they criticised the United States because of its engagement in Vietnam. Starting point of this upheaval was West Berlin; the Free University was the center and starting point for a lot of activities and movements. The SDS, the socialistic German student union which in 1964 still was a small group, then attacked by the West German newspapers from the Axel Springer company, became influential among students. The first demonstration took place at the Free University in connection with the invitation of the philosopher Erich Kuby to lecture in 1965. The world-wide conflict of the generations erupted at the Free University and led to the political right/left polarisation. An opposition group from outside the parliament, the so-called APO, *Außerparlamentarische Opposition*, was built against the great coalition of CDU and SPD in Bonn. Between 1966 and 1969 the universities faced turbulent years. The Free University was leading the whole student movement in Germany at that time. The students' central demands called for co-determination on the basis of parity, no limitation on the duration of studies and no compulsory removal from the register of students. Political topics in their discussions were the war in Vietnam, the emergency laws in 1968, the prohibition to pursue one's profession as civil servant by government ruling, and anti-authoritarian education. The latter in Germany was scientifically founded by a professor of psychology at the Free University who practiced it in a kindergarden.

Milestones of this period were

- the death of Benno Ohnesorg by firing squad in 1967 during the anti-Shah student demonstration;
- the assassination attempt against Rudi Dutschke in 1968; he was one of leaders of the student movement. He had proclaimed the "long march through the institutions" in connection with the exclusion from a civil service profession by government ruling; and the
- freeing of Andreas Baader from prison in 1970 by Ulrike Meinhof, an assistant at the Free University who had taught controversial seminars.

The so-called Berlin model of a community of professors and students at the Free University collapsed. Between 1969 and 1972 there was a kind of academic civil war at the Free University. Because of this situation, which was intensified by a new university law in Berlin (West), many of the highly qualified professors left, causing an academic slowdown at the Free University. In 1969 Rolf Kreibich, a young assistant from the social sciences, who had graduated in physics before, became president of the Free University. The NOFU, *Notgemeinschaft für eine Freie Universität* (Emergency Action Group for a Free University), and from this the society *Bund Freiheit der Wissenschaft* (Union Freedom for Science) were founded, fighting for high scientific standards and against communist indoctrination.

In the middle of the seventies the newer generation of students brought peace to the Free University. Students now worked hard in their subjects in order to prepare themselves for finding jobs after graduation. The scientific reputation of the Free University grew once again.

After the reunification of Germany the economic difficulties caused new problems for the Free University. Having lost its particular character it now has to find its way as one of the three universities in the city, with a fourth one in the neighboring city, Potsdam, and to do so without expecting special sympathy from the population and without any support from the United States. It will continue to exist but the days where it was one of the largest universities of the country have passed. It will be one university among others.

Mathematics

At the end of the war the situation in mathematics at the *Humboldt-Universität* was the following. Of the full professors, Erhard Schmidt, Ludwig Bieberbach and Harald Geppert, only the first one remained in his position. Bieberbach lost his because of his involvement in Nazi ideology and politics. Geppert had committed suicide with his family when the Soviets occupied Berlin. The lecturers Hermann Ludwig Schmid, Kurt Schröder and Alexander Dinghas had remained in or near the city till the end of the war. They started teaching already in summer 1945 before the university was officially opened. The first two soon were appointed professors and then full professors. Dinghas who was of Greek nationality was treated

Alexander Dinghas

differently. He later got a position which was equivalent to a full professorship (*persönlicher ordentlicher Professor*). Additionally, Friedrich Neiss became an associate professor.

In 1948 Helmut Hasse, who had lost his position in Göttingen in 1945, was appointed full professor; also Heinrich Grell and Karl Schröter joined the faculty. Who of these mathematicians was willing to leave for the new Free University? Obviously, Erhard Schmidt was not. He was one of the faculty members totally loyal to his university which he had joined as full professor in 1917 as successor of H.A. Schwarz, who himself had replaced Karl Weierstrass in his position when Weierstrass retired. It seems likely that one or the other of the mathematicians was thinking about changing to the Free University. But only Dinghas secretly started negotiating with the founding committee and when he accepted a position in January 1949, it became impossible for the others to follow him.

Obviously, it was not easy to fill positions at universities in Berlin in these days. Two associate professors could be hired for mathematics at the Free University, Robert Ritter from the Technical University of Berlin and Hans-Heinrich Ostmann from Marburg. In filling the second full professorship in mathematics the faculty faced severe problems. The list for this position at the end of 1951 was: 1. Friedrich Wilhelm Levi, 2. Ludwig Bieberbach, 3. Karl Schröter.

At that time Levi was already 63 and Bieberbach almost 65 years old. Levi had lost his position at the University of Leipzig in 1935, because he was Jewish. He emigrated to India where he found a job at Calcutta University and then in 1948 at the Tata Institute in Bombay. Trying to hire him was a kind of compensation for the injustice done to him by the Nazi regime. Therefore, it looked quite illogical that the second slot was reserved for one of the main former Nazi adherents, Ludwig Bieberbach. But as it was impossible to hire Bieberbach and because of his advanced age, his placement on the list strengthened the situation for Levi. After difficult negotiations between the university and the government of Berlin, Levi got the position on October 1, 1952. Bieberbach kept a low-paid research position at the Mathematical Institute of the Free University. Karl Schröter was hired by the Humboldt University of Berlin.

Levi retired at the end of March 1956. In 1959, both Ritter and Ostmann died and the mathematical institute at the Free University had to be rebuilt again after just one decade by A. Dinghas. Unfortunately, the same situation occurred after yet another decade at the end of the sixties. This time the change was, among other reasons, due to the student uprisings and the reform of the universities.

References

[1] *Eickhoff, U., Wilker, E.*, Freie Universität Berlin, eine Hochschule zwischen Reform und Konflikt, FU-Info **29** (1973), Sonderheft.

[2] Freie Universität Berlin, 1948–1973. Hochschule im Umbruch, Ausgew. u. dokumentiert von Lönnendonker, S., Fichter, T., Staat, J. unter Mitarbeit v. Rietzschel, C., Schröder, K., T 1–6, Pressestelle FU Berlin, Berlin, 1973.

[3] Fünfzehn Jahre Freie Universität Berlin, 1948–1963, Hrsg. vom Allgemeinen Studentenausschuß der FU Berlin, FU-Spiegel, Dez. 1963.

[4] *Kutowski, G.*, Der Kampf um Berlins Universität, Veritas, Justitia, Libertas, Festschrift zur 200-Jahrfeier der Columbia University, New York, Colloquium Verlag, Berlin, 1954, 7–31.

[5] *Kutowski, G.*, Freiheit. Die Gründung der Freien Universität Berlin 1948, in: Prell, U., Winkler, L. (eds.), Die Freie Universität Berlin 1948–1988, Aussichten und Einsichten, Berlin Verlag A. Spitz, Berlin, 1989, 16–30.

[6] *Weischedel, W. (ed.)*, Ideal und Wirklichkeit einer Universität, Dokumente zur Geschichte der Friedrich-Wilhelms-Universität zu Berlin, de Gruyter Verlag, Berlin, 1960.

[7] *Lenz, M.*, Geschichte der Königlichen Friedrich-Wilhelms-Universität zu Berlin, 4 vols., Verlag der Buchhandlung des Waisenhauses, Halle, 1910–1918.

[8] *Lönnendonker, S.*, Freie Universität Berlin, Gründung einer politischen Universität, Verlag Duncker & Humblot, Berlin, 1988.

[9] *Mohrmann, W.*, Geschichte der Humboldt-Universität zu Berlin von 1956 bis zur Gegenwart, ein Überblick, Humboldt-Universität, Berlin, 1960.

[10] *Müller, M., Müller, E.E.*, "...stürmt die Festung Wissenschaft!" Die Sowjetisierung der Mitteldeutschen Universitäten seit 1945, Hrsg. vom Amt

für Gesamtdeutsche Studentenfragen des Verbandes Deutscher Studentenschaften und "colloquium", Zeitschrift der freien Studenten Berlins, Colloquium Verlag, Berlin, 1953.

[11] *Prell, U., Winkler, L. (eds.)*, Die Freie Universität Berlin 1948–1968–1988, Aussichten und Einsichten, Berlin Verlag A. Spitz, Berlin, 1989.

[12] *Rabehl, B.*, Am Ende der Utopie, Die politische Geschichte der Freien Universität Berlin, Argon Verlag, Berlin, 1988.

[13] *Schröder, K.*, 1968 und die Folgen: Studentenrevolte und Reform, FU-Info **11** (1988), Sonderheft, 19–20.

[14] *Tent, J.F.*, Freie Universität Berlin 1948–1988, Eine deutsche Hochschule im Zeitgeschehen, Colloquium Verlag, Berlin, 1988.

[15] *Tent, J.F.*, Die Entstehung der Freien Universität Berlin, Berliner wissenschaftliche Gesellschaft e.V., Berlin, 1987, 278–295.

[16] 40 Jahre Freie Universität Berlin. Die Geschichte 1948–1988. Einblicke, Ausblicke, Katalog zur Ausstellung. Hrsg. vom Präsidenten der FU Berlin, Berlin, 1988.

[17] *Walther, R.Th.*, Vor 40 Jahren, Abschied von der Berliner Universität, FU-Info **4** (1988), 13–15; Vom Esplanade nach Wannsee, FU-Info **6** (1988), 8–9; Die Freie Universität Berlin – eine Hochschulgründung im Spannungsfeld der Politik, FU-Info **11** (1988), Sonderheft, 6–11.

Mathematics at the Berlin Technische Hochschule/Technische Universität

Eberhard Knobloch

The era of the independent academies

The *Berlin Technische Hochschule* was founded in 1879 through the union of two considerably older academies, the *Bauakademie* (Building Academy, founded in 1799) and the *Gewerbeinstitut* (Vocational Training Academy, founded in 1821). It later, in 1916, incorporated the former *Bergakademie* (Mining Academy, founded in 1770). These three academies were institutions whose teachers were not expected to do scientific research as such, and the teachers offered their training courses avocationally. It was not until 1856 that the first regular professorship of mathematics was established at the *Gewerbeinstitut*, a position first occupied by Karl Weierstrass (1815–1897). Only in 1895 was such a professorship created at the *Bergakademie*, its first occupant being Fritz Kötter (1857–1912). The *Bauakademie* was never provided with a professorship in mathematics.

At all three academies mathematics was regarded as an auxiliary science from the very beginning. At the Mining Academy the teachers, who lectured before only a few students, had a high turnover rate; only Daniel Chr.L. Lehmus (1780–1863) taught uninterruptedly for a long period: 1814–1852. His main appointment was to the University of Berlin (founded in 1810). Also many other teachers of the academies were associated with the university: Friedrich H.A. Wangerin (1844–1933) was with the Mining Academy; Johann Ph. Gruson (1768–1857), Peter G.L. Dirichlet (1806–1859), Ferdinand Minding (1806–1885), and Martin Ohm (1792–1872) were with the Building Academy; Karl Weierstrass (1815–1897) was with the Vocational Training Academy; and Georg Hettner (1854–1914) with the *Technische Hochschule*.

Lehmus was a typical *Dozent* (lecturer) insofar as he taught at several higher institutions, in his case at the Mining and at the Building Academies. Many of his colleagues did so: Hugo O. Hertzer (1831–1908) taught at all three academies; Ringleb, Karl Pohlke (1810–1876), Siegfried Aronhold (1819–1884) and Ernst Kossak (1839–1892) taught at the Building and the Vocational Training Academies.

The first prominent mathematics *Dozent* at the Mining Academy was Fritz Kötter (1857–1912), who very successfully applied mathematics to technology. In 1900, he changed his position after being appointed professor at the *Technische Hochschule*. His worthy successor from 1900 to 1905 became Adolf Kneser (1862–1930), who was succeeded by Eugen Jahnke (1863–1921), after the integration of the Mining Academy into the *Technische Hochschule* in 1916.

The development of the Building Academy and of the Vocational Training Academy was characterized by increasing demands on the students. The first eminent mathematics *Dozent* at the Building Academy was Dirichlet, who taught only part-time. At this Academy, special weight was given to descriptive geometry, which for a long time was taught by the painter Pohlke, thereafter (from 1877 to 1879) by Guido Hauck (1845–1905) (who was later three times elected Rector of the *Technische Hochschule*). In 1874 Julius Weingarten (1836–1910) became professor of mechanics. His publications gave ample evidence of his capability as a geometer. He was the founder of the Berlin school of classical surface theory at the *Technische Hochschule/Universität*, whose members included Scheffers, Hessenberg, Rembs, and the *Privatdozent* (private lecturer) Karl Peter Grotemeyer, who in 1958 became professor at the *Freie Universität Berlin*.

The Vocational Training Academy required a somewhat longer time than the Building Academy before its training courses attained a reasonably high scientific level. There the first eminent mathematics *Dozent* was Karl Weierstrass, who in 1856 occupied the first mathematics chair. This chair has survived to this day; in 1864 when Weierstrass was definitely appointed full professor of mathematics at the University of Berlin he was succeeded by Aronhold. Another chair of mathematics was founded in 1869 and went to Elwin B. Christoffel (1829–1900). Christoffel remained only three years in Berlin, and in 1872, E. Kossak became his successor. K. Pohlke's successor here was his assistant H.O. Hertzer in 1865, who from 1874 onward was professor of descriptive geometry.

Mathematics at the Technische Hochschule (1879–1945)

When the *Technische Hochschule* was established in 1879 by uniting two academies, the Building Academy contributed one chair (Hauck) and the Vocational Training Academy three chairs (Hertzer, Aronhold, Kossak) to the four professorships of geometry and mathematics at the *Technische Hochschule*, which existed at the beginning. They all were assigned to the section V: General Sciences. In addition to that, there were four *Privatdozenten*: Scholz and Hamburger of the Building Academy, Reichel and Buka of the Vocational Training Academy.

Weingarten was professor of mechanics. When Kötter, professor at the then still independent Mining Academy, was appointed professor at the *Technische Hochschule*, he became professor of rational mechanics. When the Mining Academy was integrated into the *Technische Hochschule* in 1916, Jahnke went with his chair to the new Mining Department. No successor to him was appointed at the *Technische Hochschule*.

Except for the special cases mentioned, departmental terminology at the *Technische Hochschule* where mathematics chairs were settled changed seven times:
1879–1922 the *Abteilungen V*, *VI* or *VII* for General Sciences;
1922–1945 the *Fakultät I* for General Sciences;
1946–1970 the *Fakultät* for General Engineering Sciences;
since 1970 the *Fachbereich* (Department) of Mathematics.
As of 1998 mathematics is to become part of a new "Faculty for Mathematics, Physics and Chemistry."

We can clearly distinguish between a geometrical and an analytical tradition at the *Berlin Technische Hochschule*. For greater clarity it seems reasonable to consider these two traditions separately. The two chair holders in geometry, Hauck and Hertzer, taught for 26 and 28 years, respectively, at the newly founded *Technische Hochschule*. The *Privatdozenten* for geometry, Scholz and Buka, finished their activities at the *Technische Hochschule* in 1889 and 1896, respectively. Georg Scheffers, a former pupil of Sophus Lie in Leipzig, succeeded Hauck in 1907. He apparently enjoyed high esteem at the *Berlin Technische Hochschule*, to which he was elected rector in 1911. Nevertheless, almost to the end of his career, Scheffers had no institutional connection with his mathematical colleagues. When he retired in 1935, no attempt was made to appoint a successor.

Hertzer's chair in geometry, on the other hand, has been occupied continuously up until the present day. In 1907 it was filled by the synthetic geometer Stanislaus Jolles (1857–1942). A far more prominent figure was his successor and former pupil of Schwarz and Hauck, Gerhard Hessenberg (1874–1925), who was mainly interested in the foundations of geometry. Unfortunately, he died still in 1925, the very year he assumed his professorship in Berlin.

Hessenberg was succeeded by the differential geometer, Erich Salkowski (1881–1943). Though in 1944 efforts were made to appoint Wolfgang Haack (1902–1994), the appointment itself did not take place.

The two chair holders in mathematics, Aronhold and Kossak, resigned from their positions at the *Technische Hochschule* in 1883 and 1892, respectively; the two *Privatdozenten* in mathematics, Reichel and Hamburger in 1883 and 1903, respectively. In 1883, Aronhold was succeeded by Heinrich Weber (1842–1913), a former pupil of Otto Hesse, but he remained at the *Technische Hochschule* for only one year. Weber was succeeded by two former pupils of Kummer, at first by Paul du Bois Reymond (1831–1889), then by Emil Lampe (1840–1918). In 1919, Georg Hamel (1877–1954), a former pupil of Hilbert, was appointed as Lampe's successor. Hamel influenced mathematics and mechanics at the *Berlin Technische Hochschule* as no other mathematician before him did. He was an expert in both disciplines. In 1945 he was at an age to resign, but he stayed on until 1949, without teaching once at the *Technische Hochschule* after World War II.

As was the case with Aronhold's successorship, there was an interregnum of two years with regard to Kossak's chair. Wilhelm Stahl (1846–1894), a former pupil of Reyhe and Weierstrass, died two years after he had assumed the professorship. He was succeeded by another pupil of Weierstrass and Kummer, Georg Hettner (1854–1914). In 1914, Hettner's position went to Rudolf Rothe (1873–1942), a pupil of Schwarz and Frobenius. During the two world wars, he became heavily involved in ballistic studies.

When Rothe resigned in 1939 his chair was occupied by Werner Schmeidler (1890–1969), a pupil of Landau, who was an expert in abstract algebra as well in applied mathematics. Schmeidler had been a national socialist since 1937. By 1945, six other pure or applied mathematicians among his colleagues had joined the National Socialist Party: Ulrich Graf in 1937, Karl Jaeckel in 1937, Paul Lorenz in 1937, Günther Schulz in 1937, Friedrich Tölke in 1933, Karl Wilhelm Wagner in 1933.

Georg Hamel

In 1927, a new professorship of mathematics and mathematical economics was established at the *Berlin Technische Hochschule*. The first to hold this position was F. Klein's pupil Aloys Timpe (1882–1959) from the *Berlin Landwirtschaftliche Hochschule*. He remained, however, isolated from his four other colleagues.

The *Privatdozenten* and *Dozenten* played an important role within the teaching staff, which had some features characteristic of *Technische Hochschulen*. The Building and Vocational Training Academies, it cannot be denied, eliminated the *Privatdozenten*. But their appointment did not require a special *Habilitation* procedure. Regulations for the conduct of a *Habilitation* at the *Berlin Technische Hochschule* were not introduced until 1884, and these regulations did not stipulate that the candidate held a doctorate degree, because the *Hochschule* was not yet entitled to confer such a degree. Experienced *Privatdozenten* had the possibility to become *Dozenten*, even though this group had no counterpart at the universities.

During the 66 years between 1879 and 1945, 39 *Privatdozenten* in mathematics took up their teaching duties after F. Buka, M. Hamburger, O. Reichel, and J. Scholz. In the course of time, eight of them were appointed professor of mathematics at the *Technische Hochschule*: W. Haack, G. Hessenberg, K. Jaeckel, E. Jahnke, St. Jolles, F. Kötter, R. Rothe, and E. Salkowski.

The expert in control theory W. Cauer, the expert in ballistics C. Cranz, the analyst and expert in aerodynamics R. Fuchs, the geometer U. Graf, the analyst J. Horn, the algebraist and number theorist E. Jacobsthal, the geometer and algebraist O.-H. Keller, the analyst and expert in potential analysis L. Lichtenstein, the algebraist E. Steinitz, and the practical analyst F.-A. Willers were among the better known of the other 31 *Privatdozenten*. The other *Privatdozenten* were: O. Dziobek, H. Grosse, A. Wendt, H. Servus, R. Müller, E. Haentzschel, A. Gleichen, G. Wallenberg, E. Meyer, W. v. Ignatowsky, A. Barneck, H. Lemke, P. Schafheitlin, E. Stübler, G. Grüß, G. Haenzel, M. Sadowsky, P. Lorenz, W. Magnus, G. Schulz, and K. Maruhn.

Mathematics at the Technische Universität (1945–1987)

The winter semester of 1944/45 was the last one before the end of World War II during which regular courses were offered at the *Berlin Technische Hochschule*. There were still six mathematicians at this time on the faculty. The three full professors Hamel, Schmeidler, and Timpe, the supernumerary professors Cauer and Schulz, and Schmeidler's assistant István Szabó.

Only Timpe and Szabó resumed their former activities when the school was reopended as the *Technische Universität* on April 9, 1946. New teachers were the lecturer Eduard Rembs (1890–1964), a former pupil of Study, and Ernst Mohr (1910–1989), a former pupil of Weyl, who was appointed full professor. Schmeidler was removed with all the other members of the National Socialist Party, by December 31, 1945. The *Technische Universität* wanted to have a new beginning. Along the way, however, these initial good intentions would become lost. At first, Schmeidler's professorship was occupied by Mohr, who had suffered from the National Socialists. Rembs had been prevented from habilitating in 1938 for political reasons. He became the official successor of Salkowski in 1949, the same year Wolfgang Haack, a former pupil of Haußner, was appointed to Hamel's professorship.

When Timpe resigned in 1959, two chairs were exchanged. Schmeidler was not only re-appointed to his old professorship, but he was also appointed as honorary senator of the *Technische Universität* in 1959. Mohr's chair was renamed the chair for mathematics (later on "Mathematics I"), five further chairs for mathematics were created between 1958 and the university reform of 1969. Since 1983, Mohr's chair has been occupied by the mathematical physicist Rolf Seiler. On the other hand, the three older, traditional chairs in geometry (Rembs), mathematics and mechanics (Haack), pure and applied mathematics (Schmeidler), were again occupied under the university reform, and partly even after then, by similarly specialized mathematicians (the years the chairs were held at the *Technische Universität* are written in brackets): Rembs was succeeded by the geometers Hans Robert Müller (1956–1963), Kurt Leichtweiß (1963–1970), Rolf Schneider (1970–1974), and Dirk Ferus (since 1975); Haack was succeeded by Hubert J. Weinitschke (1966–1977); Schmeidler was succeeded by Erich Kähler (1958–1964), Erhard Meister (1966–1970), and Ernst-Jochen Thiele (since 1972).

Haack had initiated a working group for the development of program-controlled computers, which was officially called the *Recheninstitut* (Calculating Institute) after 1956. The chair of the director of this institute was in 1964 renamed from *Lehrstuhl für Numerische Mathematik* to *Lehrstuhl für Angewandte und Instrumentelle Mathematik*; this chair was not reoccupied after Haack's resignation in 1968. In 1958, Chair II for mathematics was established and occupied by Günter Hellwig (1958–1966) and Christian Pommerenke (from 1967). In 1962 Chair III for mathematics was created and occupied by Karl Jaeckel (1962–1974), Götz Alefeld (1976–1981), Alfred Louis (since 1986). In 1964, Jaeckel took over the administration of the Calculating Institute from Haack. After his resignation in 1974, it became part of the Central Establishment Calculating Centre (ZRZ) of the *Technische Universität.*

In 1965, Chair IV for mathematics was created and occupied by Hans Wilhelm Knobloch (1965–1970), Friedrich Huckemann (1971–1986) and Günter Frank (since 1988). In 1964, a new chair for theoretical mechanical engineering was established at the Faculty for Mechanical Engineering and occupied by Rudolf Zurmühl (1964–1966) and Julius Albrecht (1966–1970). His successor, the functional analyst Rolf Grigorieff (from 1971) took up his activities at the department for mathematics.

A similar new assignment of a chair affected the traditional chair for rational mechanics. In 1900, it had been created as a full professorship at the department of civil engineering and had been occupied by Fritz Kötter (1900–1912), Hans J. Reißner (1913–1934), Friedrich Tölke (1937–1945), and István Szabó (1948–1975). Reißner was replaced by the member of the National Socialist Party, Tölke. Five other colleagues were also expelled or persecuted by the National Socialists: A. Barneck, R. Fuchs, E. Jacobsthal, E. Mohr, and E. Rembs. In 1977, the functional analyst Karl Heinz Förster was appointed to Kötter's professorship, which was assigned to the department of mathematics.

By 1987, thirty further professorships had been created since the implementation of the university reform. Thirty *Privatdozenten* also took up their activities between 1945 and 1987.

The mathematicians of the Technische Hochschule/Technische Universität Berlin and the Mathematical Society of Berlin

On October 31, 1901, the Mathematical Society of Berlin was founded. A. Kneser and E. Jahnke were the organizers of the founding group which consisted of 38 charter members. The new society was determined by the mathematicians of the *Technische Hochschule.* Fourteen of the charter members became *Dozenten* or professors at the *Technische Hochschule.* Up to 1945, the Society was mainly directed by mathematicians from the *Technische Hochschule.* In time it grew to be an accepted centre for mathematics in Berlin.

The Third Reich had a considerable impact on the Berlin Mathematical Society, too. In 1938, a number of its Jewish members were forced to resign, among them A. Barneck, E. Jacobsthal, and I. Schur. After 1945, it was not until 1950 that the Berlin Mathematical Society was re-established through the initiative of thirteen mathematicians who came from all three Berlin universities, colleges, and from the secondary schools: from the *Humboldt-Universität* (then, East Berlin); from the Free and Technical Universities and the Teachers' Training College of West Berlin. As a result, members of all three universities or colleges or secondary schools belonged to the Enlarged Executive Board of Directors (Executive Board and Council). The political background of the members was left out of consideration. The Executive Board adhered to a "purely scientific standpoint", though some members, like Rembs and Timpe, disagreed with this procedure and withdrew from the Society. This created room for Schmeidler to take the chair from 1954–1958, and the former active National Socialist Ludwig Bieberbach was admitted to the Society in 1951. In spite of the political situation of Berlin, the Executive Board consisted of members belonging to East and West Berlin until 1964. The political events of 1989 put an end to such difficulties.

References

[1] *Knobloch, E.*, Mathematik an der Technischen Hochschule/Technischen Universität Berlin 1770–1988, Berlin, 1998.

Mathematics in Berlin at the Humboldt University: from 1953 until now

Herbert Kurke

It is difficult to report on the most recent "history" of mathematics in Berlin for various reasons, not the least of which have to do with the fact that the events are still too close to the present and many persons involved in this history are still active or have retired only recently. The basic sources of this brief report are

- files from the Archives of the Humboldt University;
- *Personal- und Vorlesungsverzeichnisse der Humboldt-Universität* 1953–1968 (publication ceased after 1968, to be continued only just recently); and
- personal reminiscences (I studied at the Humboldt University from 1960 until 1964, and joined the staff of Humboldt University in 1972).

There were three political events which had great impact on the mathematical (and scientific) community in the eastern part of Germany: first the building of the Berlin Wall in August 1961; second, in 1968, the so-called *III. Hochschulreform* (university reform in the GDR); and third, the reunification of Germany in 1990.

The erection of the Berlin Wall resulted in a drastic separation from the mathematical community in the western hemisphere and an interruption of traditional scientific contacts, in particular with the western part of Germany. Thus for a long time after 1961, mathematics in the Humboldt University, or in general the scientific community in the GDR, was in a relatively isolated situation. The detrimental effect on mathematical life was partially moderated by scientific contacts with other socialist countries, in particular Russia with its strong school of mathematics. Later on, a minimal exchange with mathematical institutions in the West gradually became possible, but not for young mathematicians for whom such contact would perhaps have been most important. Contacts with mathematicians from the western hemisphere were, however, never completely interrupted thanks to visits paid to colleagues in the GDR. From the point of view of the geographical and political situation of Berlin, the mathematical department of the Humboldt University was surely in a particularly fortunate position as far as (short term) visitors were concerned.

The second event that had an immense influence on the entire university was the so-called *III. Hochschulreform*, which started in 1968. Like the reformation of the universities in the western part of Germany, it was also a step towards a university for the masses (*Massenuniversität*). The old structure of the traditional five faculties was dissolved and the so-called *Sektionen* were established instead, which were much more specialized.

Leo Kaloujnine, Karl Schröter

From the beginning of 1953 until 1968, mathematics at the Humboldt University had consisted of three mathematical institutes, namely

I. Mathematisches Institut: with 3 professors;

II. Mathematisches Institut: originally with 1 professor, later 3 professors;

Institut für Mathematische Logik: 1 professor.

The *I. Mathematisches Institut* mainly concentrated on pure mathematics, with the professors Heinrich Grell (algebra), Hans Reichardt (number theory, differential geometry), until 1956 Leo Kaloujnine (algebra), since 1960 Udo Pirl (complex functions), and since 1967 Lothar Budach. Leo Kaloujnine returned to his native country, Russia, after he held the position of a *Dozent* at the Humboldt University in 1951, and *Professor mit Lehrauftrag* (1952–1956), respectively.

The *II. Mathematisches Institut* was the institute of applied mathematics with the professors Kurt Schröder (in 1953 also Paul Lorenz and Werner Kramer until their retirement, as *Professoren mit Lehrauftrag*). From 1958 Erna Weber (statistics) was a *Professor mit Lehrauftrag* till her retirement in 1966. From 1960, until his moving to West Germany in 1968, Rolf Reissig had the chair for Analysis, which Arno Langenbach has occupied since 1966.

The *Institut für Mathematische Logik* had one professor, Karl Schröter.

A further institute was founded in 1963 and called the *Institut für Schulmathematik*. At first, it was endowed with only one chair (filled by Alfred Schroeter); one year later, a second chair was added (Klaus Härtig). This institute was mainly concerned with "Didactics of Mathematics" and with the education of teachers for mathematics.

One special institution was founded in 1966, the so-called *Spezialklassen für Mathematik und Physik*. This was an institution comparable to a *Gymnasium*, but

Hans Reichardt

attached to the university. It covered the full program of ordinary schools leading to the *Abitur* (the usual final examination in Germany, after 12, resp. 13 years of school). But it was unusual in its special emphasis on teaching mathematics (resp. physics) supervised by university teachers. It aimed at especially talented pupils, who were recommended by their school teachers and who had passed an entrance examination for this type of school.

All these institutes formed the *Fachschaft Mathematik* and were part of the traditional *Mathematisch-Naturwissenschaftliche Fakultät*, presided by the Dean of the Faculty.

The *III. Hochschulreform* mentioned above changed this structure; the role of the faculty, which still continued to exist, was drastically reduced, the new administrative units were the *Sektionen*, the *Fachschaft Mathematik* turned into the *Sektion Mathematik* and the structure of the institutes was dissolved. Instead of the former institutes, smaller units were founded and called *Bereiche*. Mathematics at the Humboldt University consisted then of the following *Bereiche*:

1. Mathematical Logic, Foundations of Mathematics, Cybernetics,
2. Algebra,
3. Geometry,
4. Analysis (for a while two *Bereiche*, Analysis I and Analysis II, were under discussion),

Heinrich Grell

5. Mathematical and Statistical Methodes of Operations Research ,
6. Mathematical Methods of Data Management, and
7. School Mathematics.

In my opinion this resulted, at least temporarily, in a rather unnatural specialization and separation among the various fields of mathematics.

There were also changes in the curricula for students, at least tentatively, towards giving special training for application-oriented work in contrast to a broader general education. Thus, so-called *Hauptausbildungsrichtungen* were proclaimed, encompassing fields four through seven on the above list, whereas fields more oriented towards pure mathematics were only considered for recruiting the staff for teaching. Most students were directed to the *Hauptausbildungsrichtungen* and as far as I remember, in the late 1960s it was even under discussion to make *Analysis* a *Hauptausbildungsrichtung*.

Education was to follow a program that was mandatory for all universities and had to be approved by the Ministry of Education. A more severe effect of the *III. Hochschulreform* was the growing political influcence of the SED (the *Sozialistische Einheitspartei Deutschlands*, the party of the communists), and the growing political indoctrination to follow the ideology of the SED. In principle the SED was very distrustful and suspicious towards people, and was obsessed with having everything under its control. Thus, for example, the following rules were fixed in the Statues (*Arbeitsordnung der Sektion*):

> The chairman of the *Bereich* is responsible for the political-ideological and the scientific direction of the *Bereich* and responsible towards the chairman of the *Sektion*.

Regularly, once a month, the chairmen of the *Bereiche* had a meeting with the chairman of the *Sektion*, where also the representatives of the SED, the FDJ (the socialist youth organization) and the FDGB (trade unions) participated. In practice, there were no democratic decisions but only instructions from the chairman of the *Sektion* (who usually was a faithful member of the SED) or the representative of the party. Enormous pressure was exerted to indoctrinate the students in the policy of the SED, and to follow faithfully all requirements (for example, to volunteer for military service).

As a result of the above-mentioned distrustful attitude of the leading clique of the SED, it became more and more difficult to get a promotion without being a faithful member of the party. There were only few exceptions.

Nevertheless, there was an active mathematical life in teaching and research, in pure and applied mathematics. In pure mathematics, there were active groups in the fields of mathematical logic, algebraic geometry and commutative algebra, complex functions (complex functions in one variable until 1988, when Udo Pirl died; in 1988 a chair was created for complex functions in several variables), differential geometry and global analysis. In applied mathematics there were active groups in applied analysis, numerical analysis, stochastics (mathematical statistics, probability, applied probability) and optimization. The formation of the group of optimization was the consequence of several developments: From the winter term 1967 until 1986 Frantiček Nožička was visiting professor at the mathematics departement; it was his scientific influence that formed this group. Moreover, in the late sixties and seventies there was great political optimism that through methods of "operations research" one should have all means at hand to solve problems of the socialist economy. This atmosphere made it relatively easy to obtain the means for creating new positions.

When looking at this picture, one notices that number theory, a domain which was traditionally very strong at the Humboldt University, no longer seems to be present. Till his retirement in 1973, Hans Reichardt concentrated his activities more on the mathematical institute of the Academy of Sciences. However, the group of number theorists of this institute regularly gave lectures at the Humboldt University and there was an actíve interchange and many common activities between the group of algebraic geometry of the university and the group of number theory at the academy.

At first, the algebra group (*Bereich* "Algebra") also did research work in general algebra; however, these activities were later on concentrated in a new *Bereich* "Computer Science" (*Informatik*) with parts of the *Bereiche* "Mathematical Logic" and "Mathematical Methods of Data Management," which eventually led to a new *Sektion Informatik* in 1988 (which, after 1990, became the *Institut für Informatik* of the university).

By 1989/90 the *Sektion Mathematik* allowed for the following scientific positions: 13 *Ordinarien* (full professors) and 15 *Dozenten* (lecturers), as well as 98 teaching assistants. There were 775 students of mathematics (*Diplom*) and school teachers of mathematics enrolled at this time.

The above-mentioned *Spezialklassen für Mathematik* still existed, as well as another interesting working body, the so-called *Mathematische Schülergesellschaft* which was founded in 1970. This association still exists. It is an organization that pupils from schools in Berlin or the vicinity can join, after having passed an entrance examination. They meet once a week in small working groups, supervised by members of the teaching staff, to get special training in mathematics and work on interesting exercises. Basically, pupils joining this association consider mathematics their hobby. Currently, about 350 pupils are members of this association.

The period after 1989/90: The autumn of 1989 and the following period of political changes and the reunification of Germany were again events with great impact on mathematics at the Humboldt University. In mathematics the change went comparatively smoothly. After all, there had been nothing like "communist mathematics," and the curriculum for the education of students was much the same as in the western part of Germany, and in research the scientific contacts had never been completely interrupted. But nevertheless, before 1989, the GDR was an almost hermetically closed region with only a few universities (hardly any new universities had been created) and, accordingly, the mobility of the scientific staff was not very high. As a result there was a big accumulation of teaching assistants with permanent positions, many of them were comparatively old and they did not properly fit into the new frame, due to cutbacks in the number of positions. Furthermore, involvement in the former political establishment and abuse of power caused many discussions. Several commissions were formed to investigate such cases.

In Germany, with its federal system, the procedure of transition to the corresponding body of law for universities (*Landeshochschulgesetz*) was slighty different in the individual states. In Berlin, on the basis of the so-called *Hochschulüberleitungsgesetz* special commissions of experts were established for each individual scientific domain, called the *Struktur- und Berufungskommission* (SBK). It was their duty to work out proposals for the structure of the newly established departments, proposals for the chairs of professors and proposals for the reengagement of the former staff.

In an exception to the practice in Germany, also the former members of the scientific staff were allowed to apply for the newly announced chairs (in general, in Germany, one cannot apply for a chair at the university where one is working). In many cases former professors of the Humboldt University were reappointed, but this was not invariably the case.

One reason behind this system was, of course, to free an academic career from dependence on the individual involvement or noninvolvement in the political system. Thus it was hoped that those who had experienced discrimination for political reasons would get a new chance at an academic career. Clearly, this was a very difficult and subtle process, as I know from my own experience as a member of such an SBK. The SBK for mathematics was established in November 1991 and finished its work in March 1994. It consisted of 3 professors of mathematics of the Humboldt University (Uwe Küchler, Herbert Kurke, Arno Langenbach), 3 external professors of mathematics (Jochen Brüning, Hans Föllmer, Klaus Hulek; later Friedrich Hirzebruch and Karl-Heinz Hoffmann), and one representative each of the teaching assistants and of the students of the Humboldt University.

The working frame for the new department of mathematics was prescribed by the government of Berlin as follows: 26 positions for professors; 49 positions for teaching assistants (unfortunately, this has been reduced to 20 professors and 41 teaching assistants in the meantime).

The following chairs were proposed, most of them equipped with a C4 and a C3 position and (with the exception of number VI which is still vacant) filled according to the proposals of the SBK.

I. *Logik*

II. *Algebra und Zahlentheorie*

III. *Algebraische Geometrie*

IV. *Differentialgeometrie und Globale Analysis*

V. *Komplexe Analysis*

VI. *Mathematische Physik*

VII. *Analysis*

VIII. *Angewandte Analysis*

IX. *Numerische Mathematik*

X. *Leiter Weierstrass Institut für Angewandte Analysis und Stochastik* (WIAS, see [3])

XI. *Optimierung*

XII. *Wahrscheinlichkeitstheorie*

XIII. *Mathematische Statistik*

XIV. *Mathematik und ihre Didaktik*

As a result, 15 former professors of the Humboldt University were reappointed in the new department as professors; for 11 positions external persons were appointed, 4 of them from the former mathematical institute of the Academy of Sciences. For the teaching assistants the situation was similar. Some of the former professors retired and some were appointed to other mathematical institutes.

The research of the institute is partially involved in several *Sonderforschungsbereiche* (supported by the *Deutsche Forschungsgemeinschaft* DFG) and *Graduiertenkollegs* (also supported by the DFG). It also incorporates the *Mathematische*

Schüler-Gesellschaft and the *Spezialklasse für Mathematik* which was re-established in a modified form. Finally, there is an active interrelation with the WIAS; the director of the WIAS is also professor at the Humboldt University. And until 1997 there was a research group "Algebraic Geometry and Number Theory" supported by a special program of the Max-Planck-Society for research groups of former institutes of the Academy of Science.

References

[1] Archive of the Humboldt-Universität.

[2] Internal Report of the SBK Mathematik.

[3] *Sprekels, J.*, Weierstrass Institute for Applied Analysis and Stochastics (WIAS), in this volume.

The Mathematical Institute of the Academy of Sciences of the GDR

Helmut Koch

Introduction

Our description of the Mathematical Institute of the Academy of Sciences of the GDR will be rather formal. We describe the development of an institution which had very few scientific collaborators at its foundation in 1946, while the number of collaborators in 1990 had grown to almost 200.

We will say very little about the scientific profile of the institute since this profile is determined by the leading mathematicians of about 20 scientific groups working in various domains; not all can be considered in the space available here, and it would not be justified to pick out the work of some and ignore the work of others.

Our report draws primarily on the *Jahrbücher der Deutschen Akademie der Wissenschaften* [1]. Up to 1966 they present detailed information including the names of all scientific collaborators. The information becomes scantier in the following years and almost disappears by 1970, the year of the "Reform of the Academy." Later, the data restricts itself to the name of the director, his deputy and the heads of the four departments (*Bereichsleiter*), with research highlights listed in just one short sentence per department; and to the publication list of the collaborators. For a report about the political atmosphere in the last years of the institute, see [2].

Foundation of the Institute

World War II brought the work of the Prussian Academy of Sciences to a standstill.

The reopening of the Academy under the new name *Deutsche Akademie der Wissenschaften* in August 1946 was therefore the start of a new institution, whose purpose was not only to continue the work of the former Academy but also to create research institutes which were to take over traditions of the *Kaiser-Wilhelm-Gesellschaft* and other governmental and even private institutions.

The *Forschungsinstitut für Mathematik* (Research Institute of Mathematics) was founded one month later on October 1, 1946. In the scope of the new institute three commissions of the Prussian Academy of Sciences continued to exist: the *Jahrbuch über die Fortschritte der Mathematik*, the *Mathematisches Wörterbuch*, and the *Edition of the Collected Works of Weierstrass*. In fact, only one of the three commissions continued to work and eventually produced, in 1961, a mathematical dictionary in the German language edited by J. Naas and H.L. Schmid. A second review journal, the *Zentralblatt für Mathematik* (see [4]), which had been in existence since 1929 parallel to the *Jahrbuch*, resumed its work in the

Kurt Schröder

Forschungsinstitut für Mathematik. Besides this editorial project, the task of the institute consisted in "pure research on all domains of mathematics." In its first three years the *Jahrbuch der Akademie* mentioned publications by the following mathematicians: E. Schmidt, G. Hamel, H. Hasse, H.L. Schmid, F. Bachmann, O. Grün.

There were six further scientists at the institute, four of them working for the *Zentralblatt*, the other two were K. Heegner and P. Roquette (for E. Schmidt, see [3], and for H. Hasse, H.L. Schmid, O. Grün and P. Roquette, see [5]).

K. Heegner later became famous for his research on elliptic curves and quadratic number fields. His paper "Diophantine analysis and modular functions" appeared in 1952 in the journal *Mathematische Zeitschrift*. Heegner died in 1965. His research work remained unknown until 1967 when H.M. Stark referred to the article in connection with his determination of all imaginary quadratic number fields with class number one.

The Institute up to 1959

In 1947, the institute founded its own mathematical journal, *Mathematische Nachrichten*. The editorial office of this journal was located in the institute until 1992. In 1949, K. Schröder became a member of the directory of the institute and developed the research in applied mathematics. In 1950, the institute was divided into two sections: *Mathematik und Editionen* (Mathematics and Editing) and *Angewandte*

Mathematik (Applied Mathematics) under the management of H.L. Schmid and K. Schröder.

H.L. Schmid, who had been the driving force in the development of mathematics in Berlin after the war, left the city in 1953 for Würzburg. His successor as the director of the section *Mathematik und Editionen* was J. Naas, who has played an important role in the development of the Academy after the war as its first director. H. Hasse, who had left Berlin already in 1950 for Hamburg, continued to be a member of the directory of the institute until 1959. The departure of these two eminent mathematicians, and of almost all their students, signified a serious loss for mathematics in the GDR.

Beginning in 1955, the directory consisted of mathematicians who were members of the *Deutsche Akademie der Wissenschaften*: E. Schmidt, H. Hasse, K. Schröder, E. Hölder and E. Kähler. In 1958 the number of scientific collaborators in the section *Mathematik und Editionen* was 19; in *Angewandte Mathematik* there were 31.

Two Institutes from 1959 to 1970

In 1959, the *Forschungsinstitut für Mathematik* was closed and two new institutes were founded: The *Institut für Angewandte Mathematik und Mechanik* (Institute for Applied Mathematics and Mechanics) with K. Schröder as director and the *Institut für Reine Mathematik* (Institute for Pure Mathematics) governed by a board of directors. This directory consisted of the heads of the five sections of the institute: H. Grell (Algebra), H. Reichardt (Number Theory), W. Rinow (Topology), J. Naas (Differential Geometry), and A. Papapetrou (Mathematical Physics).

K. Schröder, H. Grell and H. Reichardt were full professors at the Humboldt University of Berlin, and W. Rinow at the University of Greifswald. On the one hand this enabled part of their collaborators to teach at these universities besides pursuing the scientific work in the institute. On the other hand, the professors understood their role as heads of research groups to be essentially that of organizers. Their collaborators had to acquire their scientific qualification either by means of the study of the literature or/and by the contact to mathematicians outside the GDR. Since contacts to Western countries and to West Germany were difficult for political reasons, the most important scientific connections were developed to mathematical centers of the Soviet Union and also to centers in Poland and Hungary. For the number theory section the connection with I.R. Shafarevich and his group in the mathematical institute of the Academy of Sciences became very fruitful.

At the beginning, the Institute of Applied Mathematics and Mechanics consisted of the following sections: computing center, gas dynamics, ship-hydrodynamics, nonlinear mechanics and control, laboratory for photoelasticity, research on turbulence and boundary layers.

In 1963, the Institute of Pure Mathematics was extended by a section *Grundlagen der Mathematik und Mathematische Logik* (Foundation of Mathematics and

Mathematical Logic) with K. Schröter as the director. In the same year H.-J. Treder became head of the section *Mathematische Physik*, but in 1967 this section was transferred to the *Institut für Astrophysik* (Institute for Astrophysics).

Reunification of the Institute, 1971–1991

In 1969 a "reform" of the Academy began parallel to a reform of the universities in the GDR. These reforms were directed towards a closer connection of research with the needs of the industrial production in the GDR and to a stronger control of the scientific and educational activities by the communist party (SED). As a result of this reform the two mathematical institutes of the Academy of Sciences were united into a new institute originally called *Zentralinstitut für Mathematik und Mechanik* and later (in 1981), after mathematics and mechanics were separated, *Institut für Mathematik*. Finally, in 1985, the institute was named *Karl-Weierstrass-Institut für Mathematik*. The first director of this institute was K. Schröder. After his retirement in 1972, K. Matthes became director and remained in this position until the Academy was closed at the end of 1991.

Beginning with 1981 the institute was divided into four sections (*Bereiche*): *Bereich I*: Pure Mathematics, *Bereich II*: Analysis, *Bereich III*: Stochastics, *Bereich IV*: Computing Center.

The last years of the Institute

According to the Treaty of German Reunification of 1990 the institutes of the Academy of Sciences of the GDR had to be closed. A commission was established by the West German *Wissenschaftsrat* (science council), which consisted of mostly West German mathematicians, for the evaluation of the institute. The atmosphere during the period of evaluation was friendly. There was one evaluator for each group of mathematicians in the institute and mostly both sides knew each other already for some time. As a result of this evaluation, the commission proposed the foundation of a new and smaller institute for applied analysis and stochastics (see [6]). All groups working in pure mathematics were positively evaluated as well and it was proposed to transfer them to universities. This turned out to be difficult, since the universities had to reduce their staff, and a mere transferring of personnel would have been in contradiction with the principles of competition determining scientists for positions in universities. Nevertheless, the great majority of habilitated collaborators at the institute found employment in universities as associate (C3) or full (C4) professors or other acceptable positions.

We conclude this article with the list of former collaborators of the institute who were appointed professors to universities during the years 1988 to 1997:

Andreas Baudisch, Humboldt-Universität Berlin

Hellmut Baumgärtel, Universität Potsdam

Michael Demuth, Universität Clausthal

Thomas Fiedler, Université de Toulouse

Jens Franke, Universität Bonn
Jürgen Gärtner, Technische Universität Berlin
Konrad Gröger, Humboldt-Universität Berlin
Stephan Heinrich, Universität Kaiserslautern
Rolf-Peter Holzapfel, Humboldt-Universität Berlin
Burglind Juhl-Jöricke, Uppsala University
Helmut Koch, Humboldt-Universität Berlin
Henning Läuter, Universität Potsdam
Jürgen Läuter, Universität Magdeburg
Werner Müller, Universität Bonn
Reinhard Pöschel, Technische Universität Dresden
Bert-Wolfgang Schulze, Universität Potsdam
Detlef Seese, Universität Karlsruhe
Ludwig Staiger, Universität Halle
Ralph Stöhr, University of Manchester
Helge Toutenburg, Universität München
Yuri Tschinkel, University of Illinois at Chicago
Stephan Waack, Universität Göttingen
Manfred Wollenberg, Universität Leipzig
Ernst-Wilhelm Zink, Humboldt-Universität Berlin
Thomas Zink, Universität Bielefeld

References

[1] Jahrbücher der Deutschen Akademie der Wissenschaften 1946–1991, Akademie Verlag, Berlin.

[2] *Koch, H.*, Mathematik in der DDR und Wiedervereinigung, in: Wissenschaften und Wiedervereinigung, Akademie Verlag, Berlin, 1998.

[3] *Baumgärtel, H.*, Erhard Schmidt, John von Neumann, in this volume.

[4] *Ett, W., Welk. R.*, Zentralblatt für Mathematik und ihre Grenzgebiete, in this volume.

[5] *Jehne, W., Lamprecht, E.*, Helmut Hasse, Hermann Ludwid Schmid and their students in Berlin, in this volume.

[6] *Sprekels, J.*, Weierstrass Institute for Applied Analysis and Stochastics (WIAS), in this volume.

Fast Algorithms, Fast Computers: The Konrad-Zuse-Zentrum Berlin (ZIB)

Peter Deuflhard

The Konrad-Zuse-Zentrum Berlin (ZIB) is an inter-university research institute of the state of Berlin, which operates in Mathematics and Information Technology. It was founded in 1984 as an institution incorporated under public law and was named after Konrad Zuse, a former structural engineer who constructed the first programmable computer in the world. In his work, Zuse combined the thinking of both mathematics and computer science, which has become the guiding principle of the Center.

The main emphasis of research and development at ZIB is in Scientific Computing (both numerical methods and discrete methods) – an interdisciplinary field, which covers the theoretical analysis of mathematical models describing complex scientific, technical, social, and economic processes or phenomena; the development of efficient algorithms for the numerical simulation or optimization of such models; and the production of fast and robust code based on these algorithms.

The application oriented research done at ZIB contributed to solving crucial problems in science, technology, environment, and society, and in particular those problems that cannot be solved by traditional methods, but which are accessible to mathematical analysis. ZIB's part in this effort is to develop innovative algorithms and to use high-performance computers in close cooperation with researchers in science and industry.

Joint projects are being conducted at ZIB in the following fields: telecommunication, medical sciences, public transportation as well as the chemical, electrical, computer, and automotive industries.

In addition to its research branch (about 100 out of roughly 160 personnel), the Konrad-Zuse-Zentrum Berlin operates high-performance computers (currently CRAY supercomputers) as a service to universities and scientific institutes in Berlin, with about 40 personnel. The capacity of these computers is made also available to users from other parts of the country including representatives from industry and joint projects with ZIB.

The Center is well-known for its highly efficient algorithms and software, for its competence in running supercomputers, its integration into German and worldwide networks, and for its model cooperation between mathematicians, scientists and engineers. In order to make the developed algorithms at ZIB available to the scientific community, software libraries are maintained, which are generally accessible free of charge via electronic mail, the World Wide Web and other electronic media. ZIB's eLib (electronic library) runs a European node of netlib and the REDUCE Network Library. Software for the documentation of museum inventories are also available. ZIB participates in the development of internet-oriented information

Konrad Zuse

and communication systems for the sciences and is involved in the design of search engines for libraries. Preprints (for future publication in journals) and technical reports (manuals, scripts) are published. Those interested may request copies of these publications electronically via http://www.zib.de/ZIBbib or by mail. ZIB's scientific work is documented by external lectures and at workshops, conferences, presentations, and training programs held at ZIB. Finally, ZIB scientists act as co-editors of various international journals.

Weierstrass Institute for Applied Analysis and Stochastics (WIAS)

Jürgen Sprekels

WIAS was founded in 1992 out of the former mathematical institute of the Academy of Sciences of the GDR. The institute is a Leibniz-Institute, i.e. one of the 82 German research institutes organized in the so-called Research Society Gottfried Wilhelm Leibniz. WIAS is financed in equal parts by the Federal Ministry of Education, Science, Research and Technology (BMBF) and the Federal State of Berlin. Its location is in the vicinity of the *Gendarmenmarkt* in the very center of Berlin.

WIAS carries out application-oriented mathematical research projects in various fields of application, with a special emphasis on semiconductor device simulation, nanoelectronics, optoelectronics, phase transitions, stochastics in the engineering sciences and economics, and continuum mechanics. The research activities range from mathematical modelling over theoretical mathematical analysis of the models, to the development of algorithms and software for the numerical simulation of real-world technological processes.

The institute has 83 permanent positions, including 54 for scientists. It is divided into 7 research groups:

- **Partial Differential Equations and Variational Equations**
  Head: Herbert Gajewski
  The group is working on the analytical theory of such equations (existence, uniqueness, qualitative behaviour) and on the development and implementation of algorithms for their numerical solution.
  The algorithms are used for the numerical simulation in industrial applications. The main topics of research are mathematical models of carrier transport in semiconductors, reaction-diffusion equations for the transport of dopants in solids, polymerization processes, phase transitions and the heat treatment of steel.

- **Dynamical Systems and Control Theory**
  Head: Klaus R. Schneider
  The research in this group is devoted to the study of the behaviour of dynamical systems depending on parameters, with an emphasis on special structures, such as symmetries and singular perturbations. One aim is to obtain qualitative results concerning the long time behaviour, another the development of algorithms for the numerical analysis of large-scale systems and of boundary-value problems on infinite time intervals.
  The methods are applied to simulations and to the control of processes in chemical engineering, optoelectronics and problems in geophysics.

- **Numerical Mathematics and Scientific Computing**
  Acting Head: Friedrich Grund
  The work of this research group is determined by the interaction of solving real-life problems in cooperation with external users and the development and analysis of innovative numerical methods. Activities concentrate on adaptive methods for partial differential equations, large differential-algebraic systems and nonlinear optimization.
  Currently, the group is carrying out projects in multi-phase flows and reactive flow in porous media, large-scale chemical computing in process engineering and the simulation of microwave circuits.

- **Integral Equations and Pseudodifferential Equations**
  Head: Siegfried Prössdorf
  This research group deals with integral equation and boundary element methods for modelling and solving complex problems in the natural and engineering sciences. The investigations include the analytical theory of integral equations and boundary value problems with singularities, the development and theoretical analysis of numerical methods and the implementation of efficient algorithms in industrial applications.
  The research focuses on direct and inverse problems in elasticity, fracture mechanics, diffractive optics, airfoil theory, hydrology and seismology.

- **Interacting Random System**
  Head: Anton Bovier
  The scientific interest of this research group is focused on phenomena based on microscopic interactions of stochastic particle systems and the influence of random media.
  The investigations include modelling of interacting particle systems, the theoretical analysis by means of mathematical methods of statistical physics and the development of efficient methods of stochastic numerics (Monte-Carlo-simulation) for the solution of kinetic equations.

- **Stochastic Algorithms and Nonparametric Statistics**
  Head: Michael Nussbaum
  This group concentrates on algorithmic aspects of stochastics, i.e. on the development, the theoretical foundation and evaluation, and the practical implementation of statistical and numerical methods. This includes the solution of inverse and ill-posed problems by methods of non-parametric statistics, the theory and numerics of stochastic differential equations and the efficiency of stochastic algorithms.
  Special topics under study are inverse problems of image processing and tomography, methods of optimal statistical curve estimation by means of wavelets, the numerical solution of stochastic differential equations, stochastic algorithms for the numerical solution of partial differential equations as well as problems of adaptive cluster analysis.

- **Continuum Mechanics**
  Head: Krzysztof Wilmański
  The group works on problems of continuum mechanics and thermodynamics of multicomponent bodies. Special emphasis is laid on the propagation of sound and shock waves in porous and granular materials, quasistatic processes with free boundaries in porous bodies with mass exchange, and hysteresis.
  The results are applied to non-destructive testing, medical diagnosis, contamination transport in soils, crystal growth, superalloys and phase transformations.

Zentralblatt für Mathematik und ihre Grenzgebiete

Walter Ett and Reiner Welk

Abstracting and reviewing in mathematics dates back to 1871, when the first volume of *Jahrbuch über die Fortschritte der Mathematik* appeared in Berlin. The first editors were Carl Ohrtmann and Felix Müller. The publishing house used was that of Georg Reimer, which later (in 1922) became part of the Walter de Gruyter Publishing Group. In this first volume the mathematical literature (500 documents) of the year 1868 were reviewed. Subsequent editors were Max Henoch (1886–1890), Emil Lampe (1886–1918), and Leon Lichtenstein (1918–1927), respectively. In 1927 the Prussian Academy of Sciences took over the editorship, and Georg Feigl became the managing editor.

The continuously growing number of mathematical papers to be reviewed caused a continuously growing delay in publishing the yearbook with the reviews. For example, volume 51, reviewing the mathematical literature of the year 1925, was not published until 1932.

Therefore, in 1930, mathematicians from Göttingen, in particular Richard Courant and Otto Neugebauer, together with the Springer-Verlag founded the *Zentralblatt für Mathematik und ihre Grenzgebiete*, a new reviewing journal, which consciously abandoned the tie to the publishing year: the review should be published quickly after reception. The single issues of *Zentralblatt* appeared every three or four weeks. The growth of literature at that time amounted to 5000 documents per year.

Otto Neugebauer served as Editor-in-Chief even after he had to leave Germany. He continued to work in Copenhagen until he moved to the USA in 1938. His successor in office became Egon Ullrich (Giessen). The technical editorial staff worked at the Springer-Verlag, while the scientific editorial staff consisted of a few collaborators from several German universities.

With the beginning of World War II the difficulties in reviewing increased enormously, so that in 1940 the editorial staffs of *Jahrbuch* and *Zentralblatt* were fused: Harald Geppert served as joint editor, in collaboration with the Prussian Academy of Sciences, until all activities were terminated in 1944 by the disruptive conditions of the war.

After the war, in 1947, Hermann Ludwig Schmid began to reactivate the editorial work of *Zentralblatt* in cooperation with Springer-Verlag and the German Academy of Sciences (the successor of the Prussian Academy). A Research Institute of Mathematics was founded at this academy through his doing. Hermann Ludwig Schmid, who had been a member of the editorial staff of *Zentralblatt* since 1938, established the new editorial staff at this institute. The first post-war volume, No. 29, appeared in 1947; about 5000 documents per year were reviewed at that time.

The editorial staff remained in Berlin after H.L. Schmid's move to Würzburg in 1953. After his death in 1956, Erika Pannwitz, who was a member of the editorial staff (1930–1940) of the *Jahrbuch*, became Editor-in-Chief of the *Zentralblatt.*

The political division of Berlin (August 13, 1961) caused increasing difficulties for the editorial activities. At the end of 1966 a treaty between the German Academy of Sciences in Berlin, the Heidelberg Academy of Sciences, and Springer-Verlag ensured a certain amount of cooperation. As a consequence, two editorial staffs (each belonging to one of the academies) cooperated, mutual visits of the two Editors-in-Chief and the visit of a "liaison editor" by the West Berlin editorial group once a week to the East Berlin editorial group guaranteed a minimum of collaboration. Walter Romberg, connected with the *Zentralblatt* since his student days, became the Editor-in-Chief of the East Berlin editorial group. He remained in this position until the activities of the East Berlin group were abandoned by a government decision in 1977.

Erika Pannwitz resigned as Editor-in-Chief in 1969. Her successor was Ulrich Güntzer (1969–1974). Since 1974 Bernd Wegner has been the Editor-in-Chief.

In 1979 the *Zentralblatt* became part of the recently founded *Fachinformationszentrum Karlsruhe* (FIZ). The editorial staff remained in Berlin, forming the Berlin Department of FIZ.

The objective of the reviewing task changed somewhat: The electronic form of *Zentralblatt* became more and more important. First the database on STN International, later the CD-ROM version and now the new WWW-Database gained importance for the information needs of the mathematical community.

In the MATH database, mathematical literature back to 1931 (encompassing about 1,675,000 items) is available including the most recent current contents of all journals. In the printed version each year 25 volumes are issued, including an alphabetical and a key index volume, offering more than 50,000 abstracts and reviews (www.emis.de/MATH/).

References

[1] *Siegmund-Schultze, R.*, Mathematische Berichterstattung in Hitlerdeutschland – der Niedergang des "Jahrbuchs über die Fortschritte der Mathematik", Studien zur Wissenschafts-, Sozial- und Bildungsgeschichte der Mathematik 9, Vandenhoeck & Ruprecht, Göttingen, 1993.

Index

Sources of photographs

Bildarchiv Preußischer Kulturbesitz Berlin:
Friedrich–Wilhelms–University, cover
Leibniz, 2
Lagrange, 3
Weierstrass, cover, 17
Kronecker, cover, 15
Palace of Prince Heinrich, 18
Kummer, cover, 16, 62
v. Humboldt, 29
Dirichlet, 34
Kronecker, 65
Weierstrass, 72
Frobenius, 84

Staatliche Museen Preußischer Kulturbesitz, Nationalgalerie, Berlin, Bildersammlung Wilhelm Hensel (38/14), Photo Jörg P. Anders, Berlin:
Rebecca Dirichlet, 34
Death-picture Dirichlet, 36
Jacobi, 42

Archiv der Akademie der Wissenschaften, Berlin:
Crelle, 28

Hochschularchiv FU Berlin: Sonstige Photosammlung:
Opening ceremony of the university in 1946, 138

Acta Mathematica, Table Générale de Tomes 1–35 (1913):
Frobenius, cover
Schottky, 81
Kovalevskaia, 76
Fuchs, 77
Vahlen, 132
Landau, 129

Journal für die reine and angewandte Mathematik:
Vol. 157 (1927): Steiner, cover, 13
Vol. 158 (1927): Dirichlet, 14
Vol. 158 (1927): Jacobi, 10
Vol. 158 (1927): Eisenstein, 56
Vol. 158 (1927): Schwarz, cover, 79

Mathematische Nachrichten:
Vol. 4 (1950/51): Schmid, 145
Vol. 8 (1952): Hamel, 164

Skripta Mathematica:
Vol. 4 (1936): v. Neumann, 101

Sitzungsberichte der Berliner Mathematischen Gesellschaft:
Vol. 11 (1912): Euler, 4

C. Carathéodory, Gesammelte Mathematische Schriften 3, C.H. Beck'sche Verlagsbuchhandlung , München, 1955:
Carathéodory, 106

H. Freudenthal, Berlin 1923–30, Studienerinnerungen, de Gruyter Verlag, Berlin, New York, 1987:
v. Mises, 112

H. Hasse, Mathematische Abhandlungen 1, de Gruyter Verlag, Berlin, New York, 1975:
Hasse, 144

A. Pais, Einstein lived here. Claredon Press, Oxford, 1994:
Einstein, 118

AdN Zentralbild/Archiv:
Grell, 172
Schröder, 178

Presse–Bild–Zentrale, Berlin:
Dinghas, 158

Gerda Schimpf Foto, Berlin:
Schmidt, 98

I. Mathematisches Institut FU Berlin:
Steiner, 50

Private collection of H. Begehr, Berlin:
Mathematical Institute of FU at Hüttenweg, 152

Private collection of U. Bieberbach, Oberaudorf:
Bieberbach, 130

Private collection of H. Rohrbach (†), Mainz:
Schur, cover, 85

Private collection of A. Schmidt, Berlin:
Schmidt, 22

Privat collection of V. Thiele, born Reichardt, Berlin:
Reichardt–Hasse, 139
Kaloujnine–Schröter, 170
Reichardt, 171

http://www.zib.de/prospect/zuse/index.de.html:
Zuse, 184

Addresses of authors and editors

Hellmut Baumgärtel
Universität Potsdam
Institut für Mathematik
Postfach 60 15 53
D–14415 Potsdam
baumg@rz.uni-potsdam.de

Heinrich Begehr
Freie Universität Berlin
I. Mathematisches Institut
Arnimallee 3
D-14195 Berlin
begehr@math.fu-berlin.de

Reinhard Bölling
Universität Potsdam
Institut für Mathematik
Postfach 60 15 53
D-14415 Potsdam

Peter Deuflhard
Konrad-Zuse-Zentrum für
Informationstechnik Berlin (ZIB)
Takustr. 7
D-14195 Berlin
URL: http://www.zib.de

Harold M. Edwards
Courant Institute of
Mathematical Sciences
New York University
251 Mercer St.
New York, NY 10012
hme1@scires.acf.nyu.edu

Walter Ett
Zentralblatt für Mathematik
Franklinstr. 11
D-10587 Berlin

Hans Föllmer
Humboldt-Universität zu Berlin
Institut für Mathematik
Unter den Linden 6
D-10099 Berlin
foellmer@mathematik.hu-berlin.de

Jean-Pierre Friedelmeyer
I.R.E.M.
Université Louis Pasteur
7, rue René Descartes
F-67084 Strasbourg Cedex

Ralf Haubrich
Institut für Wissenschaftsgeschichte
Humboldtallee 11
D-37073 Göttingen
rhaubri@gwdg.de

Wolfram Jehne
Universität Köln
Mathematisches Institut
Weyertal 86-90
D-50931 Köln

Eberhard Knobloch
Technische Universität Berlin
Institut für Philosophie,
Wissenschaftstheorie,
Wissenschafts- und Technikgeschichte
Ernst-Reuter-Platz 7
D-10587 Berlin

Helmut Koch
Humboldt-Universität zu Berlin
Institut für Mathematik
Lehrstuhl Zahlentheorie
Jägerstr. 10-11
D-10117 Berlin
koch@mathematik.hu-berlin.de

Jürg Kramer
Humboldt-Universität zu Berlin
Institut für Mathematik
Unter den Linden 6
D-10099 Berlin
kramer@mathematik.hu-berlin.de

Uwe Küchler
Humboldt-Universität zu Berlin
Institut für Mathematik
Unter den Linden 6
D-10099 Berlin
kuechler@mathematik.hu-berlin.de

Herbert Kurke
Humboldt-Universität zu Berlin
Institut für Mathematik
Unter den Linden 6
D-10099 Berlin
kurke@mathematik.hu-berlin.de

Erich Lamprecht
Universität des Saarlandes
Fachbereich Mathematik
Postfach 15 11 50
D-66041 Saarbrücken
lamp@math.uni-sb.de

Hanfried Lenz
Freie Universität Berlin
II. Mathematisches Institut
Arnimallee 3
D-14195 Berlin

Herbert Pieper
Berlin-Brandenburgische
Akademie der Wissenschaften
Jägerstr. 22-23
D-10117 Berlin
pieper@bbaw.de

David E. Rowe
Johannes Gutenberg-Universität Mainz
Fachbereich Mathematik
Staudinger Weg 9
D-55099 Mainz
rowe@mat.mathematik.uni-mainz.de

Norbert Schappacher
UFR. Mathématique et Informatique
Université Louis Pasteur
7, rue René Descartes
F-67084 Strasbourg Cedex
schappa@math.u-strasbg.fr

Reinhard Siegmund-Schultze
Kastanienallee 12 (Haus C)
D-10435 Berlin

Jürgen Sprekels
Weierstrass Institut für Angewandte
Analysis und Stochastik (WIAS)
Mohrenstr. 39
D-10117 Berlin
sprekels@wias-berlin.de

Ernst-Jochen Thiele
Technische Universität Berlin
Fachbereich Mathematik
Straße des 17. Juni 135
D-10623 Berlin
thiele@math.tu-berlin.de

Reiner Welk
Zentralblatt für Mathematik
Franklinstr. 11
D-10587 Berlin

Vita Mathematica

Herausgegeben von
Emil A. Fellmann, Basel, Schweiz

Die Reihe ***Vita Mathematica*** *bringt unter einheitlichen Gesichtspunkten verfasste Werkbiographien bedeutender Mathematiker von der Antike bis in unsere Zeit – unter Berücksichtigung der wissenschaftshistorischen Forschung der letzten Jahrzehnte. Diese Bücher sind nicht primär für professionelle Mathematikhistoriker geschrieben, sondern wenden sich an Studierende der Mathematik, der Physik und der technischen Wissenschaften in den ersten bis mittleren Semestern, an Mathematik- und Physiklehrer, Mathematiker und Physiker, die ihre Fachdisziplinen in ihrer Einbettung in die Kultur- und Geistesgeschichte kennenlernen und studieren möchten. Die Publikationssprachen sind Deutsch, Englisch oder Französisch.*

VM 1 **Georg Cantor,** von W. Purkert und H.-J. Ilgauds (deutsch)
1987, ISBN 3-7643-1770-1

VM 2 **Blaise Pascal,** von Hans Loeffel (deutsch)
1987, ISBN 3-7643-1840-6

VM 3 **Heinrich Heesch**, von Hans-Günther Bigalke (deutsch)
1988, ISBN 3-7643-1954-2

VM 4 **George Berkeley**, von Wolfgang Breidert (deutsch)
1989, ISBN 3-7643-2236-5

VM 5 **Norbert Wiener**, von Pesi R. Masani (englisch)
1990, ISBN 3-7643-2246-2

VM 6 **André Weil**,
Souvenirs d'apprentissage, Autobiographie (französisch)
1991, ISBN 3-7643-2500-3

VM 7 **Johannes Faulhaber**, von Ivo Schneider (deutsch)
1993, ISBN 3-7643-2919-X

VM 8 **Christian Goldbach**, von A.P. Juškevič und Ju.Kh. Kopelevič (deutsch)
1993, ISBN 3-7643-2678-6

VM 9 **Friedrich Wilhelm Bessel**, von K.K. Lawrinovič (deutsch)
1994, ISBN 3-7643-5113-6

VM 10 **Bernhard Riemann**, von B. Laugwitz (deutsch)
1995, ISBN 3-7643-5189-6

VM 11 **Evariste Galois** von L. Toti Rigatelli (englisch)
1996. ISBN 3-7643-5410-0

SN • Science Networks • Historical Studies

Edited by
Erwin Hiebert, Harvard University, Cambridge, MA, USA
Hans Wussing, Universität Leipzig, Germany
in cooperation with an international editorial board

The publications in this series are limited to the fields of mathematics, physics, astronomy, physical chemistry, and their applications. The publication languages are English preferentially, German, and in exceptional cases also French. The series is primarily designed to publish monographs. However, special editions featuring collected letters as well as thematic groupings of smaller individual works or proceedings can be taken into consideration. Annotated sources and exceptional biographies might be accepted in rare cases. The series is aimed primarily at historians of science and libraries; it should also appeal to interested specialists, students, and diploma and doctoral candidates. In cooperation with their international editorial board, the editors hope to place a unique publication at the disposal of science historians throughout the world.

SN 15 Ch. Sasaki / M. Sugiura / J.W. Dauben (Eds)
The Intersection of History and Mathematics
1994 (ISBN 3-7643-5029-6)

SN 16 I. Yavetz
From Obscurity to Enigma - The Work of Oliver Heaviside, 1872–1891
1995 (ISBN 3-7643-5180-2)

SN 17 L. Corry
Modern Algebra and the Rise of Mathematical Structures
1996 (ISBN 3-7643- 5311-2)

SN 18 K. Hentschel
Physics and National Socialism - An Anthology of Primary Sources
1996 (ISBN 3-7643-5312-0)

SN 19 D. Ullmann
Chladni und die Entwicklung der Akustik, 1750 - 1860
1996 (ISBN 3-7643-5398-8)

SN 20 I. Grattan-Guinness / G. Bornet (Eds)
George Boole - Selected Manuscripts on Logic and Its Philosophy
1997 (ISBN 3-7643-5456-9)

SN 21 J. Sakarovitch
Épures d'architecture
De la coupe des pierres à la géométrie descriptive XVIe–XIXe siècles
1998. (ISBN 3-7643-5701-0)

SN 22 P. Marage / G. Wallenborn
The Solvay Councils and the Birth of Modern Physics
1998 (ISBN 3-7643-5705-3)